TELESCOPES, TIDES, *and* TACTICS

Stillman Drake

TELESCOPES,

TIDES, *and* TACTICS

A Galilean Dialogue about
the *Starry Messenger* and
Systems of the World

The University of Chicago / *Chicago and London*

The University of Chicago Press, Chicago 60637
The University of Chicago Press, Ltd., London

© 1983 by The University of Chicago
All rights reserved. Published 1983
Printed in the United States of America
5 4 3 2 1   83 84 85 86 87 88 89

Library of Congress Cataloging in Publication Data

Drake, Stillman.
   A Galilean dialogue about the Starry messenger and
systems of the world.

   Incorporates a translation of Galileo's Sidereus
nuncius.
   Includes bibliographical references.
   Includes index.
   1. Galilei, Galileo, 1564–1642. Sidereus nuncius.
2. Astronomy—Early Works to 1800.   3. Satellites—
Jupiter—Early works to 1800.   I. Galelei, Galileo, 1564–
1642. Sidereus nuncius. English. 1983.   I. Title.
QB41.G178D7   1983   523.9'85   82-24790
ISBN 0-226-16231-1

STILLMAN DRAKE is emeritus professor of
the history of science at the University
of Toronto and author of many books on
Galileo, including *Cause, Experiment, and
Science*.

Contents

In affectionate memory of
MARIA LUISA BONELLI-RIGHINI
Director of the Institute and
Museum of the History of Science
at Florence,
where in her gracious spirit
Galileo lived again
for Italian and foreign visitors alike

Preface

This book presents Galileo's *Starry Messenger* of 1610 in the context of an imagined dialogue among three of his friends late in 1613. It is an astronomical companion to my *Cause, Experiment, and Science*, in which Galileo's first book on physics was similarly presented. Emphasis here is on Galileo as a working astronomer during the early days of the telescope. In that respect this book supplements *Galileo At Work* with new information drawn from letters and notebooks.

Telescopes transformed astronomy during the seventeenth century. Academic attempts to exclude new celestial discoveries from science as mere optical illusions had first to be overcome. In that process Galileo made his telescope an instrument of unprecedentedly precise measurement as well as one of observation and discovery. The story, recaptured from his working papers, unfolds here in discussions among close personal friends who knew Galileo's work in both physics and astronomy. It is a story of early modern science challenging traditional metaphysical accounts of the universe.

My purpose is primarily to show that the actual work of a pioneer scientist is hardly less interesting than the more usual scholarly analyses of its philosophical implications. All passages translated from Galileo's writings are denoted by a star in the margin and lines separating them from my fanciful dialogue reconstructing his work as a physicist and as an astronomer.

The interlocutors in this imagined dialogue differ from those in the previous one (and in Galileo's own later dialogues) by omission of Simplicio as spokesman for university professors of natural philosophy. In his place I have put a celebrated theologian and amateur of science, Fra Paolo Sarpi. Salviati's task is to inform the two Venetians of Galileo's work at Florence from 1610 to 1613. In return, Sagredo can acquaint him with work done by Galileo during his years at Padua, except for that done after mid-1608, when Sagredo was sent to Aleppo for three years as Venetian consul to Syria. Simplicio, who in my previous dialogue was Cesare Cremonini, ranking professor of philosophy at Padua, would not have known at first hand of Galileo's early telescopic work. Sarpi, on the other hand, was personally involved not only in encouraging that work, but in Galileo's successful presentation of his telescope to the Doge of Venice. Moreover, it was to Sarpi that Galileo first wrote of his discovery of the law of falling bodies, in 1604, and it is in Sarpi's notebooks that we find the first summary of the tide theory espoused by Galileo.

Another purpose of this book is implied in its title. Galileo's explanation of tides by Copernican motions of the earth is associated with his first expression of preference for the new astronomy, to Kepler in 1597, with the famous *Dialogue* for which he was tried and condemned in 1633, and with events in 1615 and 1616 that affected Galileo's tactics in his campaign for freedom of scientific inquiry. At the end of 1613, when my imagined dialogue was held at Venice, there had been as yet no overt intervention by the Church in the Copernican issue. Previous treatments of Galileo with respect to that issue center on the place of the Church in it. This book attempts to draw together various threads of Galileo's Copernicanism when he had in hand all the important scientific evidence he was ever to have but was not yet constrained to adapt his tactics of presentation to any arbitrary ruling by the Church such as was issued in 1616.

In his *Starry Messenger* Galileo promised soon to write a book on the system of the world. During the next six years no religious prohibition existed to deter him, yet Galileo did not fulfill his promise. Readers are invited to reflect on what else may have held him back and to reconsider the common opinion that he was headstrong and impetuous by nature, especially in supporting the Copernican cause.

Source material for the conversations and translations presented here is contained primarily in the most complete edition of Galileo's works, edited under the direction of Antonio Favaro. Existing English translations are cited in the margins of this book by page numbers preceded by a key identified below. Original language references are indicated by volume (small Roman numeral) and page in *Opere di Galileo Galilei, Edizione Nazionale* (Florence: G. Barbèra, 1929–39); such references are valid also for other printings of that work. Page references in the margins of the principal translation are to volume iii, part 1 of the *Opere*.

CES   S. Drake, *Cause, Experiment, and Science* (Chicago: University of Chicago Press, 1981).

D   Galileo, *Dialogue Concerning the Two Chief World Systems*, trans. S. Drake (Berkeley and Los Angeles: University of California Press, 1953, 1967).

D&O   S. Drake, *Discoveries and Opinions of Galileo* (New York: Doubleday Anchor Books, 1957).

GAP   S. Drake, *Galileo Against the Philosophers* (Los Angeles: Zeitlin & Ver Brugge, 1976).

GAW   S. Drake, *Galileo At Work* (Chicago: University of Chicago Press, 1978).

GS   S. Drake, *Galileo Studies* (Ann Arbor: University of Michigan Press, 1970).

KC   E. Rosen, *Kepler's Conversation with Galileo's "Sidereal Messenger"* (New York and London: Johnson Reprint, 1965).

M&M   I. E. Drabkin and S. Drake, *Galileo "On Motion" and "On Mechanics"* (Madison: University of Wisconsin Press, 1960).

Listed below are books and articles in English, mainly of recent date, from which more detailed information can be obtained concerning matters discussed in the dialogue and other points of view than mine are represented. The listing is in order of publication, those of more recent date being generally based on more detailed studies of original documents.

General Background

J. L. E. Dreyer, *A History of Astronomy* (New York: Dover Books, 1953).

A. R. Hall, *The Scientific Revolution* (London: Longmans, Green & Co., 1954).

L. Geymonat, *Galileo Galilei*, trans. S. Drake (New York: McGraw-Hill Book Co., 1965).

A. Van Helden, *The Invention of the Telescope* (Philadelphia: American Philosophical Society, 1977).

R. K. DeKosky, *Knowledge and Cosmos* (Washington: University Press of America, 1979).

Galileo and the Telescope

G. Righini, "New Light on Galileo's Lunar Observations," (with commentaries by O. Gingerich and W. Hartner), in *Reason, Experiment, and Mysticism*, ed. M. L. Bonelli and W. R. Shea (New York: Science History Publications, 1975), pp. 59–94.

S. Drake, "Galileo's First Telescopic Observations," *Journal of the History of Astronomy* 7 (1976): 153–68.

E. A. Whitaker, "Galileo's Lunar Observations," *Journal of the History of Astronomy* 9 (1978): 155–69.

S. Drake, "Galileo and Satellite Prediction," *Journal of the History of Astronomy* 10 (1979): 75–95.

K. Sakurai, "The Solar Activity in the Time of Galileo," *Journal of the History of Astronomy* 11 (1980): 164–73.

S. Drake and C. Kowal, "Galileo's Sighting of Neptune," *Scientific American* (Dec. 1980): 74–78.

Galileo and Tides

E. J. Aiton, "Galileo's Theory of the Tides," *Annals of Science* 10 (1948): 44–57.

S. Drake, "Origin and Fate of Galileo's Theory of Tides," *Physis* 3 (1961): 185–94.

H. L. Burstyn, "Galileo's Attempt to Prove That the Earth Moves," *Isis* 53 (1962): 161–85.

E. J. Aiton, "Galileo and the Theory of the Tides," *Isis* 56 (1965): 56–61.

Galileo and Tactics

G. de Santillana, *The Crime of Galileo* (Chicago: University of Chicago Press, 1955).

A. Koestler, *The Sleepwalkers* (London: Hutchison, 1959).

J. J. Langford, *Galileo, Science, and the Church* (Ann Arbor: University of Michigan Press, 1966).

W. R. Shea, *Galileo's Intellectual Revolution* (London and New York: Macmillan, 1972).

P. Feyerabend, *Against Method* (London: NLB and Atlantic Highlands, Humanities Press, 1975)

A. Koyré, *Galileo Studies*, trans. J. Mepham (Hassocks: Harvester Press, 1978).

W. Wisan et al., *New Perspectives on Galileo*, ed. R. Butts and J. Pitt (Dordrecht and Boston: D. Reidel Publishing Co., 1978).

M. L. Finocchiaro, *Galileo and the Art of Reasoning* (Dordrecht and Boston: D. Reidel Publishing Co., 1980).

I am grateful to Doubleday Anchor Books for permission to reuse my previous abridged translation of Galileo's *Starry Messenger*, first printed in *Discoveries and Opinions of Galileo*. It is here completed by inclusion of nightly observations of Jupiter's satellites from January to March, 1610. Printed diagrams accompanying those in the original edition in Latin were not entirely faithful to a set of drawings made by Galileo, almost certainly for use by the printer at Venice, which are preserved with his working papers at Florence. The drawings have been used here and are reproduced in reduced facsimile at the beginning of the Appendix, showing the scale of "minutes" used by Galileo. Numerical indications of apparent magnitudes given for the final two weeks have been removed from reproductions in the text, which does not allude to them.

Corrections and suggestions offered by my friends Noel Swerdlow and James H. MacLachlan, who read the typescript, are gratefully acknowledged. I am again indebted to Beverley Jahnke for turning my drafts into proper typescript. And above all I thank Florence Selvin Drake for her patience and encouragement during my long preoccupation with Salviati, Sagredo, Sarpi, and Simplicio.

Introduction

The telescopic discoveries announced in the *Starry Messenger* aroused great interest and excitement outside the universities. A century of novelties following the discovery of America had accustomed the public to the fact that there were more things on earth than had been dreamt of by philosophers, and it seemed not impossible that the same might apply to the heavens. Telescopes powerful enough to reveal Jupiter's satellites remained few for several months; even Johann Kepler, the greatest living astronomer and the founder of modern optics, appealed to Galileo to send him one.

Kepler did not doubt Galileo's assertions; he urged a search for satellites around other planets. Within the scholarly community generally the reaction to the *Starry Messenger* was very different. Professors of philosophy, far from demanding telescopes, rejected Galileo's announcements as the product of optical illusions if not simply a hoax. The ranking philosophers at Galileo's two universities, Pisa and Padua, refused even to look through his telescopes. They knew by pure logic, as did the leading astronomer at Rome, that what Galileo had seen in the heavens could not actually be there. A group was assembled by the professor of astronomy at Bologna to look through Galileo's own telescope; having seen the points of light near Jupiter, all denied that those were, or could be, starlets circling that planet. One young foreign astronomer wrote to Kepler of this and was the first to publish a book rejecting the discovery. At Florence another young man, who had likewise seen the objects through Galileo's instrument, published another and longer book "arguing the new planets out of the sky," as Galileo put it.

Reception of Galileo's first astronomical book paralleled that of his first book on physics two years later. In existing university science there was no place for cosmological novelties, though outside the universities, even among men who had been educated in them, newly observed objects and Galileo's ideas about them aroused interest and even excitement. In contrast with modern scholarly and public responses to new discoveries these facts deserve examination, particularly because Galileo himself was a recognized member of the scholarly community as professor of mathematics at the University of Padua.

Science was dominated by the natural philosophy of Aristotle in all universities almost from their inception four centuries earlier. Aristotle defined science as the understanding of things in terms of their causes, which are revealed by reasoning, not directly by experience. For celestial bodies and motions, causes had been laid down by Aristotle in his *De caelo*. In the universities of Galileo's time there was no higher court of appeal.

The cosmology of *De caelo*, for metaphysical reasons, presented the earth as immovable at the center of the universe. Stars and planets, as well as the sun and moon, circled around it at fixed distances from the earth and at uniform speeds. To account for observed irregularities of apparent motions, Aristotle's *Metaphysics* described an ingenious scheme of interconnected spheres carrying the planets, sun, and moon, the axis of one uniformly rotating sphere being held at a fixed angle in another.

About a century after Aristotle's death, astronomers employing careful measurements of angles and of times as a basis for predicting future planetary positions and eclipses of the sun or moon had to depart from this scheme. Hipparchus found that either the earth was not at the center of the sun's motion or that the sun had a second motion also around another center. The same was true of the moon and of each planet, so Greek astronomers introduced eccentric spheres and epicyclic motions in their work of predicting positions and eclipses.

Natural philosophers had metaphysical and physical reasons for preserving the earth at the center of the universe, against the findings of astronomers. To prevent conflict, an astronomer contemporary with Hipparchus proposed an admirable suggestion. More than a century before the Christian era, Geminus remarked that as mathematicians, astronomers did not need to concern themselves with causes and were even better off in some matters if they could ignore them. They would therefore abstain from speaking of physics, leaving that to philosophers trained in causal analysis. Thus astronomers first embarked on exact science in the modern sense by giving up science in Aristotle's sense. Cosmology remained aloof by the simple expedient of declaring eccentric spheres and epicyclic motions to be mathematical fictions. True and real celestial motions, philosophically speaking, were still perfectly uniform and circular about the earth as their common center. The task of astronomers was strictly to "save the phenomena," as actual appearances were called in Greek.

Some three centuries later Ptolemy succeeded in completing a handbook of mathematical astronomy, called the *Almagest*, that adequately served for predictions within the range of accuracy permitted by naked-eye instruments. Ptolemy stated at the outset that only on the mathematical theory of celestial motions could agreement be expected among philosophers, because the inconstancy of matter prevented their agreement in physical theory, and the intangibility of objects of metaphysical inquiry rendered agreement among philosophers there unattainable.

Mathematical astronomy was not neglected in the universities, but it was not usually taught by professors of natural philosophy, who normally confined themselves to cosmology and physics. Ptolemaic astronomy was taught by professors of mathematics, who traditionally left physics to the natural philosophers. After 1543 a few professors of mathematics also taught Copernican astronomy, but none in Italy at Galileo's time. It had no advantage for the kind of calculations most students would ever make, which were chiefly for astrological purposes and habitually used standard Ptolemaic tables of motions.

One might suppose that just as Geminus had offered a working basis that avoided conflict between philosophers and astronomers in placement of the earth, the same or a very similar agreement would have been possible in the matter of the earth's motions assumed by Copernicus. The difficulty was that Aristotelian physics required the assumption of an immovable earth, toward which heavy bodies naturally fell in straight lines. That physics was not disturbed by assuming the earth to be elsewhere than at the center of the universe, but it could not accommodate an earth in motion, let alone in rotation at something like a thousand miles an hour. Tycho Brahe avoided this new difficulty by a different kind of suggestion from that of Geminus. Tycho expected astronomers to accept Aristotelian physics, denying Copernican motions to the earth while accepting the Copernican assumption that all planetary motions were centered near the sun and very far from the earth.

It is easy to see why natural philosophers did not welcome the proposal of Tycho as they had, since antiquity, approved that of Geminus. It was gracious of astronomers to adhere to Aristotelian physics, but Tycho's price for this was too high; natural philosophers would have to alter their entire cosmology. Professors of philosophy generally considered the right of mathematicians to interfere in true science to have been yielded once and for all in antiquity, as if astronomers must remain forever content to deal only with fictions invented to fit mere appearances.

Galileo, like Kepler, knew that mathematically there was no difference between the astronomies of Copernicus and Tycho, while for ordinary practical purposes the Ptolemaic astronomy served as well as either of them. But they also knew, as everybody knows (because it is built into the very language of science, Aristotelian or modern), that among the various possible arrangements and motions for the earth and the heavenly bodies only one choice can be correct. Whether it is possible to decide which is the correct choice was another matter. If that were easy to decide, men versed in science would not have disputed over the merits of three astronomies. The right choice was difficult, but not necessarily impossible.

Astronomical considerations alone had not settled the question. Galileo himself still accepted the traditional arrangement, with the earth motionless at the center of the universe, when he moved from Pisa to Padua late in 1592. But in 1597 he wrote to Kepler that he was a Copernican and had been for several years. He said further that he preferred not to make his opinion known publicly. That remained true until his first telescopic discoveries a dozen years later. Not before 1613 did Galileo publish a flat statement that he believed the earth in motion, even if he hinted at this three years earlier in the *Starry Messenger*.

When Galileo finally published a discussion of rival world systems, in 1632, his book had been first organized as a dialogue on the tides, though (for reasons explained in the Epilogue) it was not printed with that title. Until a few years ago it was supposed that Galileo's explanation of tides in terms of the earth's motions did not occur to him much before 1610, when he first mentioned in a letter that he had composed a treatise on the subject. But in 1597, writing to Kepler, Galileo said that by the earth's motions he had been able to explain some natural phenomena (that is, some physical events) that could perhaps not be explained without them. Kepler correctly guessed that Galileo alluded to the tides, but since Kepler did not believe that they could be explained in such a way he did not ask for further information.

An outline of Galileo's tide theory is found among notes of Fra Paolo Sarpi written in 1595, within three years after Galileo's move to

Padua. Tides are virtually nonexistent in Italy except at Venice, where they rise some 5 feet. The enormous weight of so vast a quantity of water could hardly fail to induce Galileo, whose interest up to this time had been chiefly in motion and mechanics, to seek an explanation of its periodic shifting up and down. His original idea, which he went on elaborating later, is described in this book. No trace has been found of any previous speculations by Galileo requiring the Copernican astronomy, so it is probable that his initial preference for it was inspired by this problem in terrestrial physics.

Before the telescope, Galileo had no other evidence of his own as a basis for supporting Copernican motions of the earth and no further mention of them is found in his correspondence or working papers until 1610. Late in that year he observed the phases of Venus and promptly wrote a letter saying that Kepler and other Copernicans had philosophized correctly. The phases of Venus rendered Aristotelian cosmology (and Ptolemaic astronomy) no longer tenable. Astronomers recognized this, though few were won over by it to Copernicanism. Tycho's astronomy enabled them to abandon Aristotelian cosmology without surrendering its companion physics. Possibly some natural philosophers did the same, but those whose writings I have seen would abandon no part of Aristotle's science.

Galileo had no higher regard for Aristotelian physics than for traditional cosmology. By the time he heard of the telescope he had virtually completed a new science of motion based on careful measurements of distances and times in the spontaneous descent of heavy bodies, a science as different from Aristotelian physics as scientific astronomy was from Aristotelian cosmology. With the advent of the telescope he turned his attention for several years to astronomical observations, measurements, and calculations. Since Galileo published little beyond some results of those activities, they escaped attention until recently.

From Galileo's working papers it is seen that early in 1612 he transformed his telescope from a mere instrument of discovery into one of measurement far more precise than had ever been possible in astronomy. This resulted in quite good tables of the satellite motions, by means of calculation from which he detected the occurrence of a satellite eclipse during one of his previous recorded observations. His method of calculation then showed him how to predict satellite eclipses. That procedure required motions of both Jupiter and the earth around the sun to be taken into account, providing Galileo with further evidence that the earth's motion was to be regarded as real in astronomy. His one positive statement to that effect in print appeared in the appendix to his *Sunspot Letters* concerning satellite predictions.

Galileo's work in the years 1610–13 sheds new light on his Copernicanism, a subject of great interest in relation to his own career and to the Scientific Revolution of the seventeenth century. Without taking his calculations into account or knowing the early date of Galileo's tide theory, it has long been customary to suppose that Galileo's Copernicanism was purely speculative in origin, reflecting only a passion for mathematical simplicity of an abstract kind. In that view Galileo's Copernicanism never had properly scientific evidence in its support. While it is true that Galileo did not make contributions to planetary astronomy of the kind to which Copernicus, Tycho, and Kepler devoted their lives, he did make measurements and calculations relating to satellites that provided him with convincing evidence in favor of the earth's motions. New telescopic discoveries confirmed physical evidence from the existence of tides in large seas, in Galileo's view, making it clear to him that the old physics must be discarded along with the old cosmology. The long-delayed wedding of physics and astronomy, banned by Aristotle in antiquity, was the eventual fruit of the Scientific Revolution. Kepler and Galileo both attempted to join them, in quite different ways, without success; Descartes tried still another approach, and Newton finally completed the marriage.

Since the time of Newton, universities have become centers of leadership in a kind of science that places the highest value on unforeseen discoveries and novel explanations. Why that was not the case before the time of Newton is best understood in terms of the drastic change that took place in the concept of science itself, pioneered by Galileo in his first books during the years 1610–13. How the new concept was grounded in observations and measurements is suggested in the ensuing imaginary dialogue.

THE FIRST DAY

Interlocutors: *Sagredo, Sarpi, Salviati*
Scene: *Sagredo's palazzo at Venice*

Sagredo   Welcome, Fra Paolo; this is an unexpected pleasure. Having been told that you were absent on a mission for the Most Serene Republic, I feared you would not return in time to meet a guest of mine from Florence who will soon be embarking for Spain. He brings news from our old friend Galileo Galilei, so I know you will wish to see and talk with him. At the moment he is strolling along the canal; let us sit in the garden and await his return, which should be in a few minutes.

Sarpi   Thank you, Sagredo. My mission was successfully completed much sooner than could have been hoped. I arrived here only yesterday and have just made my report to the Council of Ten. I then hurried here because I was told at my convent that you had left word a week or more ago that you wished to see me. Now I know why, and nothing but the most urgent duties could tear me away before I have met your guest. If he knows what our friend is now doing, there is something I should very much like to ask him about.

Sagredo   You will find him most thoroughly informed, for it was at his villa in the hills near Florence that our friend composed the two books he has published since he left Padua. On his arrival we spent some days with Professor Simplicio in a discussion of the book about bodies that sink, rise, or float in water.

Sarpi   Though my question concerns a different interest of our friend, I wish I might have been present. I must say that I am astonished to hear that the stern Aristotelian professor would spend even an hour on so mundane a subject as hydrostatics. Tell me how it happened.

Sagredo   Actually one thing led to another. It began, as I remember, with mention of scientific demonstrations, from which, as you know, Simplicio excludes mathematics. Being told that our friend now speaks of mathematics as the language of the book of Nature, he

CES 27,   challenged that metaphor. In reply, my guest showed how de-
123, 127   monstrative advance based on sensate experience had been used by our friend to extend an Archimedean science on mechanical principles taken from Aristotle. But here he is now, to speak for himself.

By great good fortune, Salviati, Fra Paolo Sarpi has returned to Venice and is here to ask news of our old friend. Fra Paolo, I present Signor Filippo Salviati, of whom I was just speaking.

Sarpi   I am delighted that you did not escape from Venice before I had the pleasure of your acquaintance, Signor Salviati.

Salviati   The hospitality of Sagredo did not permit me to, Fra Paolo. Already I have allowed one ship to sail without me, enchanted as I am by the beauties of your city. Your return here crowns that reward with another—the opportunity to convey special greetings from our friend. I am greatly honored to know you, who hold so high a place in his esteem and affection. He asked me to assure you that only the press of new duties and further investigations has been responsible for his failure to write to you, for which he begs forgiveness.

Sarpi   Please tell him that I quite understand, and add that my involvement in affairs of state would in any case have made it impossible for me to keep up my end of a serious scientific correspondence.

GS 144   It is incredible how greatly continual attention to politics has dulled any faculty I once had for such discussions of science as we used to hold when he was professor at Padua.

Sagredo   Excuse my intervening, Fra Paolo, but I must protest; Salviati shall carry no such slander from my house. Perhaps you believe a talent to be dwindling when it is but suffering from disuse. I have no doubt that a few hours in the company of Salviati, if you can spare them, will remedy that. He shares that gift our friend has, of rousing scientific talents that lie dormant, as I learned in that discussion we had with Professor Simplicio of which I told you.

Salviati   I think you flatter me, Sagredo, mistaking for an ability of mine some habits of questioning and explaining that I have imbibed in three years of close association with our friend. Be that as it may, I share your doubt that Fra Paolo can have lost in so short a time his keen flair for scientific investigations of which our friend has often spoken and of which he is the best possible judge. In any event I am at your service, Fra Paolo, if you wish to put the matter to the test.

Sarpi   Well, as I told Sagredo before you came, there is in fact a question about our friend's present activities that has for some time been uppermost in my mind, and I hastened here as soon as I had discharged my official duty hoping that his guest from Florence would be able to throw light on it. Good fortune has decreed that the guest is one specially qualified to answer it, if you will.

Salviati   As I said, I am at your service.

Sarpi   In his *Starry Messenger*, or *Message*, as some here prefer to call it,
our friend promised soon to publish a book on the system of the
world. Three years have since gone by, and two books of his have
appeared, but they are books of a totally different kind. Except

CES 18   for an apology that his promise has not yet been fulfilled, I find
no hint in them that he remembers it. My question accordingly
is whether our friend is now at work on the promised treatise or
has abandoned the idea—and if the latter, what can it be that
occupies his time that he would count as more important than
such a profound philosophical disquisition?

Salviati   Pardon me if I seem to quibble, Fra Paolo, but you said one
question was uppermost in your mind. Yet you ask at least three
questions, and in a way many more. I suspect that what you
really seek could be given only as a full account of our friend's
thoughts since he left Padua. For suppose that I reply thus: our
friend is neither at work on the promised book, nor has he abandoned it; he counts nothing as more important, but is spending
his time on matters that seem to him no less important, recognizing them as necessary preparation unsuspected when he made
the promise. Would you accept that as a proper answer?

Sarpi   Of course not; you have caught me out. But I fear that a proper
answer may impose too greatly on your time and patience, now
that you have hinted at the cause of our friend's delay. You have
illuminated that for me as a flash of lightning shows us the nature of a terrain but does not enable us to perceive essential
details.

Sagredo   Excuse me for intruding, but you two must have some kind of
information that I lack. Possibly I blinked while the lightning
was flashing; at any rate I did not see a landscape illuminated for
even a moment.

Sarpi   It is no fault of yours, Sagredo, if Salviati's reply had for me
some power to reveal the nature of our friend's present problems,
though of course not the problems themselves. Perhaps you both
might like to hear why, and then Salviati can say whether I have
guessed right.

Ten years ago our friend frequently discussed with me a number of questions about natural motions that had not been taken up by Aristotle or accurately answered by Peripatetic philosophers. In particular there were mathematical questions about heavy bodies descending freely or rolling down inclined planes. At the beginning he showed me a treatise *De motu* that he had written at Pisa long before, in which some rules had been deduced that he found on actual test not to agree with measurements. His conclusions had been carefully reasoned, and it was certain that Aristotle had been wrong on many things, but our friend decided not to publish his treatise as long as any conclusions departed from careful observations of nature. I recall that early in 1604 he discovered the rule governing those natural motions, and later the same year he sent me a proof based on speeds of impact, but then he found that although the rule agreed with exact measurements, the proof contained a false assumption. Since he has still published nothing on that subject—though the law is beyond doubt, most surprising, and remains unknown to philosophers—I suppose he remains silent because he still lacks complete scientific knowledge of natural motions.

M&M 37, 65, 69

GAW 100–103

Thus it came about that when Salviati told us that our friend has postponed his book on the system of the world while he is at work on necessary preparations, I glimpsed a parallel with his work on the science of motion. It is one thing to be sure that Aristotle and the Peripatetics are wrong but quite another to offer demonstrative proof of a new science of motion, let alone to explain the system of the world. Have I caught your hint, Salviati?

Salviati You have indeed, though until this moment I was unaware of that other work of our friend's on the science of natural motions.[1] Its existence may in fact bear on the very delay you inquire about, so please tell us about the true law of descending heavy bodies. Our friend has not mentioned it in my hearing since he came to Florence.

Sarpi The more reason for me to leave that to him, at the time he deems appropriate. No, Salviati, you are not going to escape from answering my question by enticing me to answer yours; I am here to listen to news of our friend's recent thoughts, not to speak of his previous ones.

Sagredo   In principle I agree with you, Fra Paolo, but for reasons of my own I want to see a bargain struck between you. Putting aside that question of natural motions, there are things I wish to learn from you about the early days of the telescope here and at Padua, for I was in Syria[2] at the time of our friend's first telescopic discoveries and his promise to write on the system of the world, which doubtless arose from them. You, who were here, may know things that will interest Salviati no less than myself, so you must agree not just to listen but to inform us.

Sarpi   Gladly, if Salviati will oblige me and does not think it too great an imposition on his time to explain our friend's delays.

Salviati   Rather, you must tell me how much time you can spare for this, as it might take three or four days to tell you what occurs to me concerning your inquiry. No doubt other questions will arise also, and even a day or two might exceed your freedom from official duties—to say nothing of the burden placed on our gracious host.

Sarpi   As it happens, the mission I have just completed has enabled me to dispose already of the urgent matters of state on which my opinions and advice were sought. I now am entitled to a short vacation, as it were, and would gladly devote it to this purpose. If that would inconvenience Sagredo, perhaps you could start now and we might continue on the morrow in the cloisters of my monastery. How say you, Sagredo?

Sagredo   I say that surely you are jesting, for I know you do not mean to question my hospitality. I could no more listen to the beginning of Salviati's reply and then forgo hearing the remainder than I could have read the title page of the *Starry Messenger* and then put down the book. This is the more true because what is now promised is exactly what Professor Simplicio resolutely refused to hear of during our recent colloquies. Just as he would not look through our friend's telescope three years ago, lest it disturb his philosophy, so he will now listen to no talk of conclusions our friend has drawn through further use of that remarkable instrument. Especially, he would hear nothing of that Copernican folly, as he termed it. Thus I stand to gain no less than you, Fra Paolo, and my house and garden are yours for as many sessions as you like.

Salviati   Then we agree that I should answer Fra Paolo's question about the activities which have kept our friend from publishing his system of the world, in whatever way I deem suitable with an abundance of time at my disposal. Very well; I suggest that we start as we did with Simplicio before, this time reading the text of the *Starry Messenger*. That will refresh our memories of exactly what it was that our friend hurriedly published to the world, inviting a series of attacks and controversies that consumed a good deal of the time he might otherwise have applied to his other promised book. At the same time we shall become aware of the unfinished business, so to speak, left by his hasty publication scarcely three months after the first telescopic observations of the heavens. Some of that business has since been finished, taking more of his time, and at the proper places I can tell you how it was completed. After we have reviewed the contents of the book, which, as you recall, was a short one, and have considered the controversies and further work that ensued, we may more understandingly consider the state of our friend's promised book. Does that sound satisfactory?

Sagredo   It does to me, having found a similar beginning to be congenial and profitable in the matter of physics only a few days ago. Before Fra Paolo replies, I may add that it was our practice to take turns reading the text, so that no one felt obliged to omit interruptions and questions by reason of any obligation to hold strictly to the text before him. Each of us must be free to speak out at will, and to carry any discussion as far as it may go without loss of interest to the others. Those are the informal rules of our game, Fra Paolo; do they suit you?

Sarpi   Why, I am enchanted. It is a long time since I read the *Starry Messenger* and I think that our doing this will recall to each of us the incredulity and excitement we felt when it first appeared. Moreover, this will show us how far our thoughts about astronomy have advanced during what is, in another sense, a very short time. Please have the book brought and place it first in the hands of Salviati; as a Florentine, he may most fittingly read its letter of dedication to his sage young sovereign.

Sagredo   A servant was already sent to fetch it while Salviati was still speaking, for I was sure that you would like his suggestion. While we are waiting, Fra Paolo, you should begin by setting a background for our reading. As I said, I was absent when our friend began his work that led to the book, and by the time I returned he had moved to Florence. Since neither Salviati nor I know firsthand about some previous events here at Venice, you must tell us of them.

Sarpi   Gladly. About the beginning of November, 1608, I received from Holland a little printed newsletter[3] containing an account of an instrument, devised by a spectacle-maker of Middlebourg, that made distant things appear closer by showing them as enlarged.

I wrote at once to friends abroad to know the truth of this. Long

GS 143   ago it had occurred to me that such an effect might result from a parabolic glass, but I never attempted to grind one. On hearing of the new spyglass I thought that someone abroad might have done that, but it turned out that the Flemish instrument was quite different. Among those who wrote to me was Jacques Badovere, whom Sagredo may remember as having dwelt with our friend for a time as one of his students at Padua. He informed me that the effect of enlargement was quite real and that imitations of the Flemish spyglass were already being sold at Paris, where he lives, though they were very feeble and were little more than toys.

During the summer of 1609 our friend did not visit Florence, as had become his custom, but remained at the university where one of his students was completing the doctorate and required his presence. Also, his work on the science of motion, which had begun to progress rapidly in 1607, had at last reached the point at which he began to write a formal treatise about natural motions of heavy bodies. We had often discussed that science over the many years of our acquaintance, whereas he had never shown any great

	interest in astronomy, nor was he thinking of that when he first heard reports of the Dutch spyglass.

Sagredo   From what I know of him, he was probably most taken with the possibility of gaining an advantage for Venice over the Turks by our having a spyglass useful to our navy.

Sarpi   You are very nearly right. In June he had approached the noble Signor Piero Duodo, who was visiting Padua, about an increase in salary, but the negotiations at Venice for this proved unsuccessful. Our friend first heard news of the spyglass on a brief visit to Venice in July, and he perceived that he might be able to make one having naval value to the Most Serene Republic. As soon as he heard the reports, which some believed and others ridiculed, he visited me to ask my opinion. I showed him Messer Badovere's letter attesting to the existence of the Dutch instrument, and he returned immediately to Padua to attempt its reinvention and construction in his workshop there.

Sagredo   When I returned from Syria, I heard stories that just at that time a foreigner visited Venice with one of the instruments, which he tried to sell to our government at a high price, and that he was refused. Did such a remarkable coincidence actually occur?

Sarpi   It did indeed, and by still further coincidence the foreigner arrived in Padua just after our friend had left there to visit Venice. Some at Padua saw the foreign instrument, as our friend found out when he returned, but by the same prank of fate the stranger had just left for Venice.

Sagredo   Then our friend obtained a great practical advantage, being enabled to learn from others at Padua how the instrument was made.

Sarpi   Not at all, for the owner would not permit anyone to examine his spyglass further than to peer through it. The price asked for it was a thousand ducats, so much that the senators hesitated to act without advice and appointed me to make a report. Naturally I wanted to study its construction, but was forbidden by the foreigner to open it. All I could learn was that it had two glasses, one at either end of a long tube, so that is all that our friend could have been

GS 147–48   told at Padua. It was in fact not very powerful, enlarging a distant line by only three times. Knowing from the newsletter that the Dutch already had a more powerful spyglass, I advised the senate against this expenditure of public funds, and the foreigner angrily departed.

Salviati   I am curious to know why you say the Dutch already had a more powerful spyglass.

Sarpi   The newsletter reported that from The Hague it enabled one to distinguish windows in a church at Leiden, four leagues distant, which would have required two or three times the power of the one I saw. Right at this time I received a letter from our friend saying that he had achieved the enlarging effect, though but weakly, and was confident that within a short time he would manage to make the effect much greater. I doubted that he would count an enlargement of less than triple, and years of experience had taught me that he did not make idle promises about instruments, so I felt certain that my advice to the senate was sound.

Sagredo   Did he tell you how he had found the secret at all in so short a time?

Sarpi   Not in his hurried letter, but later he said that he first reasoned that one of the two glasses must be convex and the other concave. A flat glass would have no effect; a convex one would magnify

D&O 245   objects, but not clearly and distinctly, while a concave glass would reduce the apparent size, but might perhaps remove the indistinctness. Trying two spectacle lenses, with the concave one close to his eye, he found that the effect was produced. The problem was then to grind the concave lens deeper than is done for spectacles to aid the nearsighted and to shape the convex lens to the radius of a very large sphere, increasing the effect. For obvious reasons he did the work himself, not wishing any lens-grinder to know his plan if it succeeded. It did, and in mid-August he returned to Venice with a spyglass magnifying eight times or more. With this, from the campanile at Saint Mark's, he described approaching ships two hours before they could be seen at all by trained observers.

Sagredo   We know how he then presented this to the Doge and in return received a doubled salary and life tenure in his position at the university, though he soon resigned that professorship and entered the service of Cosimo II de' Medici at the Tuscan court. Now, what made him think to apply this commercial and naval device to the purposes of astronomy?

Sarpi   The newsletter said at the end that stars invisible to the normal eye were seen through the Dutch spyglass. Perhaps our friend soon verified that, or found it out by himself. But the turn of events so busied him that I had little opportunity to talk with him before he moved back to Florence a year after the events I have just described.

Salviati   Perhaps I can shed some light on what happened next. Having presented his first spyglass to the Doge, our friend bethought himself of his duties to his natural prince and former pupil. Hastening to Florence, he displayed to Cosimo a similar instrument useful for military purposes, and it occurred to him that a still more powerful one would make a suitable gift to the young grand duke. In the same way that he had brought the three-power toy to greater strength, he intended to perfect this further. For that, however, he needed hard, clear glass of thickness not used by spectacle makers, and in order to avoid risk that others would anticipate him if his material requirements became known here, he had glass sent from Florence of the quality and in the sizes he needed. From that he ground lenses suitable for a telescope about double the power he had yet achieved, which was already almost triple the power of toys made with spectacle lenses. He completed this late in November, and while trying it out near dusk he happened to turn it to the moon, of which only a thin crescent was illuminated. Through the telescope the moon presented sights so different from anything expected, both as to its lighted and its darkened part, that for a whole month it occupied our friend's exclusive attention. What happened at the end of that time, and how stars and planets came next to attract his interest, is set forth in the book that has been placed in my hands and from which I shall now begin reading. First is this long title:

☆ [53]

STARRY MESSENGER
Proclaiming great, faraway, and admirable sights,
hinting and proposing them to everyone
but especially to philosophers and to astronomers,
which by
GALILEO GALILEI, Florentine patrician,
Professor of Mathematics at Padua,
have recently been found to be observed in
the face of the moon, innumerable fixed
stars, the Milky Way, nebulous stars, and
above all in four planets circulating around
Jupiter, at different distances, in different
periods, and with truly remarkable swiftness,
known to no one before this time
first newly discerned by the author
and decreed to be called
THE MEDICEAN STARS.
Venice, printed by Tomasso Baglioni, 1610
With permission of the authorities, and copyright.

Sagredo   I wonder if you would begin by telling us how our friend replies to people who find it abhorrent that he styled himself messenger of the stars, as if he spoke for the heavens themselves. Among divines especially I have heard that reproach. Because of it, many people here now call it the "Starry Message," instead of "messenger."

Sarpi   I recall that he decided on the title *Sidereus Nuncius* at the last moment, when the printers had finished with most of the text and

D&O 19   demanded wording for the title page and dedication. The title he had had in mind was "Astronomical Message," as will be seen on the first page of text. But it then occurred to him that what carries a message is a messenger and that a very attractive title for a book containing news from the stars would be the one chosen. Thus it was the book, and not its author, that is there called *nuncius*, or ambassador—though the same word may also mean simply *message*.

Salviati Our friend still customarily refers to the contents of the book by our Italian word for message, *avviso*. He certainly never claimed to be the messenger of the heavens, as if he were an angel. An English divine and poet has amusingly described him instead as a commander who has summoned the stars themselves to come closer and give a better account of themselves to mortals.[4]

Sagredo Considering how some Italian authors have responded, Fra Paolo, it is a good thing for our friend that his ambassador reached the British. His own countrymen impute to him mere vanity when he wanted his good fortune to benefit all mankind, and particularly philosophers, as he said on the title page.

Salviati There, having decreed the Medicean name to be that of the new-found wandering stars accompanying Jupiter, he appropriately dedicated the book to Cosimo in the following words:

☆ [55]   To the Most Serene Cosimo II d' Medici
Fourth Grand Duke of Tuscany:

Surely a distinguished public service has been rendered by those who have protected from envious hatred the noble achievements of men excelling in virtue, and have thus preserved from oblivion and neglect names that deserve immortality. So likenesses sculptured in marble or cast in bronze have been handed down to posterity; to that we owe our statues, both pedestrian and equestrian; so we have columns and pyramids whose cost (as the poet says) is astronomical; and last, so entire cities have been built to bear the names of men deemed by posterity worthy of commendation through the ages. For the nature of the human mind is such that, unless stimulated by images of things acting on it from outside, remembrance of the originals quickly passes away.

Looking to things still more stable and enduring, some have entrusted the immortal fame of illustrious men not to marble and metal, but to the custody of the Muses and to imperishable literary monuments. But why dwell on these things as if human ingenuity were satisfied with earthly regions and dared not advance farther? Seeking farther, and well understanding that all human monuments eventually perish through violence of the elements or by great age, ingenuity has in fact found monuments still more incorruptible, over which voracious time and envious ages have been unable to assert their rights. Turning heavenward, man's wit has thus inscribed on the familiar and everlasting orbs of most bright stars the names of those whose famous and godlike deeds have caused them to be accounted worthy of eternity in the companionship of the stars. Thus the fame of Jupiter, of Mars, of Mercury, Hercules, and other heroes whose names are borne by stars will not fade until the extinction of the stars themselves.

[56] Yet this invention of human ingenuity, noble and praiseworthy though it is, has been out of fashion for many centuries. Primeval heroes are in possession of those bright abodes and hold them in their own right. In vain did the piety of Augustus attempt to elect Julius Caesar into their number, for when he tried attaching the name "Julian" to a star that appeared in his time (one of those bodies that the Greeks call *comets* and the Romans also named for their hairy appearances), it vanished in a short time and mocked his too-ambitious wish. But we, most serene prince, are able to read Your Highness into the heavens far more correctly and auspiciously; for scarce have the immortal graces of your spirit begun to shine on earth when bright stars appear in the heavens as tongues to tell and celebrate your surpassing virtues for all time. Behold, then, four stars reserved to bear your famous name—bodies that belong not to the inconspicuous multitude of the fixed stars, but to the bright ranks of the planets. Variously moving around most noble Jupiter as children of his own, they complete their orbits with marvelous speed, executing at the same time, with one harmonious accord, mighty revolutions every dozen years about the center of the universe—that is, the Sun.[5]

The Maker of the stars Himself has indeed seemed by clear indications to direct me to assign to these new planets Your Highness's famous name in preference to all others. For just as these stars, like children worthy of their sire, never leave Jupiter's side through any notable distance, so—and indeed who does not know this?—clemency, kindness of heart, gentleness of manner, splendor of royal blood, nobility in public affairs, and excellency of authority and rule have all fixed their home and habitation in Your Highness. And who—I ask again—does not know that all those virtues emanate from the benign star of Jupiter, next after God the source of all good things? Jupiter; Jupiter, I say, at the instant of Your Highness's birth, having lately emerged from the turbid mists of the horizon to occupy the middle of the heavens, illuminating the eastern sky from his own royal house, looked out from that exalted throne on your auspicious birth, pouring forth all his splendor and majesty in order that your tender body and your mind (already adorned by God with most noble ornaments) should imbibe with the first breath that universal influence and power.

D&O 16

[57]

But why should I employ mere plausible arguments when I can prove my conclusion absolutely? It pleased almighty God that I should instruct Your Highness in mathematics four years ago, at that time of year when it is customary [for professors] to rest from the most exacting studies. Now, since it was clearly mine by divine will to serve Your Highness and thus to receive from near at hand the rays of your surpassing clemency and beneficence, what wonder is it that my heart is so inflamed as to think day and night of little else than how I, who am indeed your subject not only by choice but by birth and lineage, may make myself known to you as most grateful and most desirous of your glory? Therefore, most serene Cosimo, having discovered under your patronage these stars unknown to any astronomer before me, I have with good right decided to designate them by your august family name. And if I am the first to have investigated them, who can justly blame me if I also name them, calling those the Medicean stars in the hope that this name will bring them as much honor as the names of other heroes have bestowed on other stars? For—to say nothing of Your Highness's most serene ancestors, whose everlasting glory is testified to by the monuments of history—your virtue alone, most worthy Sire, can confer upon these stars a name immortal. No one can doubt that you will fulfill the expectations, high though they be, that you have aroused at the beginning of your auspicious reign, and not only meet but far surpass those. For when you have conquered your peers you may still vie with yourself, and you and your greatness will increase with every day.

Accept then, most clement prince, this noble glory reserved for you by the stars. May you long enjoy those blessings sent to you not so much by the stars as by God, their creator and their governor.

Your Highness's most devoted servant,
Galileo Galilei

Padua, March 12, 1610

Sagredo   As you may well guess, Salviati, there were many in Venice who were angry that our friend soon afterward left Padua to enter the service of that prince, who quickly showed his wisdom by employing him as philosopher and principal mathematician at the Tuscan court.

Sarpi   Cosimo's wisdom in that matter is beyond question. I find reason to doubt, however, that it will turn out to have been wise for our friend to leave the free air of our Most Serene Republic to dwell at Florence, where the power of the Jesuits is very great. There he may not find it acceptable to depart so far from Aristotle as to make the sun, rather than the earth, the center of the whole universe.

Salviati   Rumblings of trouble on that score have been heard this year, though in a different quarter. A professor of philosophy at the University of Pisa has suggested to our prince that our friend's astronomical opinions contradict Holy Scripture. But I take it, from what you have just said, that as a theologian you see no such conflict?

Sarpi   Of course not, though such a flame may be lighted and fanned by priests less concerned with the principles of Christianity than with holding power to dominate Catholic education. The Bible shows *D&O* 186   men how to go to Heaven, not how the heavens go, as Cardinal Baronius once remarked at the home of my late friend Signor Gianvincenzio Pinelli in Padua. I recall that our friend was present on that occasion, when conversation in the vivacious company that used to meet there turned to his theory that motions of the earth *GS* 201   could explain the tides, a perennial marvel that he had first been able to observe carefully here at Venice. Someone asked whether motions of the earth around a motionless sun could be reconciled

| | with the Bible, to which the eminent cardinal replied with that witty and memorable saying. It is Aristotelian natural philosophy alone that is contradicted by a moving earth, a philosophy first challenged at its roots when Copernicus worked out his new astronomy.[6] |

Salviati   I should like to hear more about that theory of the tides you mentioned as our friend's. Tides are not a common topic of discussion at Florence, and only once do I recall having heard our friend make a passing remark about them.[7] That remark has suddenly taken on a new possible significance to me. From what you have just said, it may also bear on the principal question that we are examining. I hesitate to say more, however, until you enlighten me about his theory.

Sarpi   I shall be glad to do so, but not now. First we ought to finish what we have begun, and besides, it is many years since our friend explained his idea to me. I remember that I made some notes about it at the time, and if you like I will try to find those and bring them with me after we have finished our reading. If I am not mistaken, our friend had not previously mentioned the new astronomy, and he seldom spoke of it in later years,[8] our common interest having been in terrestrial physics rather than in celestial appearances.

Salviati   Very well, I shall resume the reading, but please do not forget to bring those notes if you can find them. The text now begins with the title mentioned earlier:

☆ [59]

ASTRONOMICAL MESSAGE
Which contains and explains recent observations
made with the aid of the new spyglass
concerning the surface of the moon,
the Milky Way, nebulous stars, and
innumerable fixed stars
as well as four planets never before seen
and now named
THE MEDICEAN STARS.

Great indeed are the things which in this brief treatise I propose for observation and consideration by all students of nature. Great, I say, because of the excellence of the subject itself; the entirely unexpected and novel character of these things; and finally, by reason of the instrument by which they have been revealed to our senses.

Surely it is a great thing to increase the multitudinous host of fixed stars visible before to our unaided vision, showing them plainly to the eye in numbers ten times greater than the old, familiar stars.

It is a very beautiful thing, and most gratifying to the sight, to behold the body of the moon—distant from us some sixty terrestrial radii [here by mistake the text has "diameters"]—as if it were no farther away than two of those measures, so that its diameter looks almost thirty times enlarged, its surface nine hundred times, and its volume increased twenty-seven thousand times over what is viewed with the naked eye. In this way one may learn with the full certainty of the senses that the moon is not clothed in a smooth and polished surface but is in fact rough and uneven, everywhere covered (like the earth's surface) with huge prominences, deep valleys, and chasms.

[60]

Again, it seems to me a matter of no small importance to have ended the dispute about the Milky Way by making its nature manifest to the senses as well as to the intellect. Likewise it will be a pleasant and elegant thing to show that the nature of those stars that astronomers have until now called *nebulous* is very different from what has been previously believed. But what far surpasses all other wonders, and what particularly moves us to seek the attention of all astronomers and philosophers, is the discovery of four wandering stars not known or observed by anyone before us. Like Venus and Mercury, which have their own periods about the sun, these have theirs about a certain star, conspicuous among those already known, which they sometimes precede and sometimes follow without ever departing from it beyond certain limits.

All these facts were discovered by me not long ago with the aid of a spyglass that I devised after first having been illuminated by divine grace. Perhaps other things, still more remarkable, will be discovered by me or by other observers with the aid of such an instrument, the form and construction of which I shall first briefly explain, as well as the occasion of its having been devised. Afterward I shall recount the story of the observations I have made.

About ten months ago a report reached my ears that a certain Fleming had constructed a spyglass by means of which visible objects, though very distant from the eye of the observer, were distinctly seen as if nearby. Of this truly remarkable effect several experiences were related, to which some persons gave credence while others denied them. A few days later the report was confirmed to me in a letter from a noble Frenchman at Paris, Jacques Badovere, which caused me to apply myself wholeheartedly to investigate means by which I might arrive at the invention of a similar instrument. This I did soon afterward, my basis being the doctrine of refraction.

First I prepared a tube of lead, at the ends of which I fixed two glass lenses, both plane on one side while on the other side one was spherically convex and the other concave. Then, placing my eye near the concave lens, I perceived objects satisfactorily large and near, for they appeared three times closer and nine times larger [in area] than when seen with the naked eye alone. Next I constructed another one, more accurately, which represented objects as enlarged more than sixty times [in area]. Finally, sparing neither labor nor expense, I succeeded in constructing for myself so excellent an instrument that objects seen through it appeared nearly a thousand times larger and over thirty times closer than when regarded with our natural vision.

[61]

Sagredo   Excuse me for interrupting, but I wonder when these things were done, for I should like to know how long it took our friend to advance from a magnification of only three to ten times that much.

Sarpi   The first, capable only of triple enlargement, was made early in
D&O 244   August 1609, at which time our friend wrote hastily to tell me that he had already succeeded in producing the effect which we had discussed in Venice only a day or two before. The second, of about eight power, he brought to Venice soon after the middle of August, where he exhibited it to many senators and persons of authority who perceived its great value to our republic. On the

26th day of that month he presented it to the Doge. Salviati has already told us that by the end of November our friend had constructed a telescope of about double the previous power, but that would still fall far short of magnification 30 times. He showed me such an instrument just before the book was published here at Venice.

x, 290

Salviati   That final telescope of highest power, like the first toy of but three power, was of limited use for observations in the heavens. Along with great power went narrowing of the area observed, so that instruments of middle power turned out to be the most useful. Last year, while our friend dwelt with me at Le Selve and was improving his tables of motions of the Medicean stars, he used instruments of 18 or 20 power, which he fitted with a device for measuring separations between the starlets and Jupiter. The telescope he used at first for observing the moon was, I believe, of about 15 power, and he gave it to Cosimo, to whom he also presented his "Discoverer."[9]

Sarpi   What you say would throw light on the curious error that you mentioned a few minutes ago, where the text said "sixty terrestrial diameters" instead of sixty earthly radii. The distance of the moon is about thirty terrestrial diameters, so if seen as though no farther than two of *those* measures it would be magnified fifteen times. Perhaps our friend, having completed the final thirty-power instrument[10] before sending his book to the printer, made a mistake in altering some correct statement he had previously written for his fifteen-power telescope.

Salviati   It is certain that the mistake was inadvertent, since our friend knew the distance of the moon as determined long ago by astronomers. Your suggestion, Fra Paolo, seems probable to me, and when I return to Florence I shall ask our friend about the circumstances. Doubtless he will wish to correct the mistake in any new edition of the *Starry Messenger*. Meanwhile, to proceed:

☆   It would be superfluous to list the number and importance of the advantages of such an instrument both at sea and on land. Forsaking terrestrial observations, however, I turned to celestial ones; and first, I saw the moon from as near at hand as if it were scarcely two diameters distant.

This would confirm Fra Paolo's conjecture and my belief that the power of the telescope used for observing the moon was about fifteen.

☆   After I had observed often and with wondering delight both the planets and fixed stars and saw the latter to be very crowded, I began to see, and eventually found, a method by which I might measure their distances apart. Here it is fitting to convey certain cautions to all who intend to undertake observations of this sort. In the first place it is necessary to prepare quite a perfect telescope that will show all objects as bright, distinct, free from haziness, and magnifying them at least four hundred times and thus showing them twenty times closer. Unless the instrument is of that kind it will be in vain to attempt to see all the things that I have observed in the heavens, which will presently be set forth.

Sagredo   This is interesting. The very powerful telescope shown to Fra Paolo was not needed in order to confirm all the discoveries to be set forth, and our friend appears to have determined, before publishing them, the least power that critics would have to achieve with their instruments before they could safely deny the truth of his statements. That power was somewhat greater than our friend had achieved when he began his reported observations, according to what you have said, Salviati, but much less than he had reached before his book went to the printer.

Salviati   I believe that later on in the book we shall have some more information about these things; meanwhile, let us return to the text:

☆   Now, to determine without great trouble the magnifying power of an instrument, trace on paper the outlines of two circles (or two squares) of which one is four hundred times as large [in area] as the other, as will be the case when the diameter [or diagonal] of one is twenty times that of the other. Then, with two such figures attached to the same wall, observe them both simultaneously from a distance, looking at the smaller one through the telescope and at the larger one with the other, unaided eye. This may be done without difficulty, holding both eyes open at the same time, and the two figures will appear to be of the same size if the instrument magnifies objects in the said ratio.

Here I pause to say that the instrument later devised by our friend for measuring very accurately the distances between Jupiter and its companions was based on the same principles as this test of telescopic power. At first, however, he attempted a different method, next described, that was unreliable and occasioned some inaccuracies of which I shall speak when we come to them later on in his book.

☆   Such an instrument having been prepared, we seek a method of measuring distances of separation [between stars]. This we shall accomplish by the following contrivance.

[62]   Let ABCD be the tube and E the eye of the observer. If there were no lenses in the tube, rays would reach the object FG along straight lines ECF and EDG; but when the lenses have been inserted, the rays go along refracted lines, ECH and EDI. Thus the rays [that reach the eye] are brought closer together, and those which were formerly directed freely to [embrace] the object FG include now only the portion HI of it. The ratio of distance EH to line HI being then found, one may by means of a table of sines determine the size of the angle formed at the eye by the object HI, which we shall find to be only a few minutes of arc. If we now fit to the lens CD thin plates, some pierced with larger and some with smaller apertures, and put

now one plate and now another over the lens, as required [to fit the separation being observed], we may form at will different angles, subtending more or fewer minutes of arc, and by this means we may easily measure intervals between two stars separated by but a few minutes, with no error greater than one or two minutes.[11] For the present, let it suffice that we have touched lightly on these matters and have scarcely more than mentioned them, as on another occasion we shall explain the whole theory of this instrument.

Sagredo   Our friend has never set that theory forth, perhaps because a year later the German astronomer Johann Kepler published a book called *Dioptrics*, in which he investigated many possible combinations of lenses, creating an entire new science. I have been study-

xi, 379   ing it with the aid of our friend's successor at Padua, Professor Camillo Glorioso, but find it very difficult. So did our friend, who remarked that perhaps even its author did not fully comprehend parts of it[12]—alluding, I believe, to the gulf between such things considered mathematically and the practical problems encountered in actually constructing useful instruments.

Salviati   Perhaps so; at any rate, Kepler early implored our friend to send him one of his telescopes so that he could confirm the claimed

iii, 184   observations for himself, as he soon did by using an instrument our friend had sent to the Archbishop of Cologne.

☆   Now let us review the observations made during the past two months, once more inviting the attention of all who are eager for true philosophy to the first steps of such important reflections. We shall speak first of the surface of the moon that faces us. For greater clarity, I distinguish two parts of that surface, a lighter and a darker. The lighter part seems to surround and to pervade the whole [visible] hemisphere, while the darker part discolors the moon's surface like a kind of cloud and makes it appear covered with spots. The spots that are quite dark and rather large are evident to all and have been seen throughout the ages; these I shall call the large or *ancient* spots, thus distinguishing them from others, smaller in size but so numerous as to be found all over the lunar surface, especially in the lighter part. These latter spots had never been seen by anyone before me. From observations of them repeated many times, I have been led to the opinion and conviction that the moon's surface is not smooth, uniform, and exactly spherical, as the majority of philosophers believe it (and all other heavenly bodies) to be, but uneven, rough, and full of cavities and prominences, it being (not unlike the earth's surface) in relief with mountain chains and deep valleys. Things I have seen that enabled me to draw this conclusion are as follows.

[63]

Sagredo Of all the new conclusions set forth in the book, this one which our friend placed first was the least acceptable to philosophers, who require perfection in all the heavenly bodies. Even astronomers of high reputation rejected our friend's view. For example, Father Clavius of the Jesuit College at Rome eventually saw and granted

x, 442, 480, 484
xi, 93

the existence of the Medicean stars (after first having derided them as illusions of the lenses), but he would still not concede roughness to the moon.

Sarpi A great furor was caused by the attack leveled against our friend by Jesuits at Mantua who defended the perfectly spherical moon after

x, 460–63

a certain criticism had been urged in Germany. Some say the Italian adversaries denied outright that our friend had ever really seen the *surface* of the moon through his spyglass, though I believe they

xi, 118

alluded only to a certain theory about the moon of which I shall speak later. The lengths to which Jesuits will go to enforce their authority were illustrated by my would-be assassins a few years ago.

Salviati The lengths to which our friend's adversaries have gone is best deferred until we have read exactly what he said that so aroused them. Here he continues:

☆   On the fourth or fifth day after new moon, when the moon is seen with brilliant horns, the boundary that divides the shaded part from the lighted does not run uniformly along an oval line, as would happen on a perfectly spherical solid, but traces out an uneven, rough, and very wavy line as shown in the figure below. Indeed, many luminous projections extend beyond the boundary into the darker portion, while on the other hand some dark patches invade the illuminated part. Moreover, a great multitude of small darkened spots, entirely separated from the dark region, are scattered almost all over the area illuminated by the sun, excepting only that part occupied by the large and ancient spots.

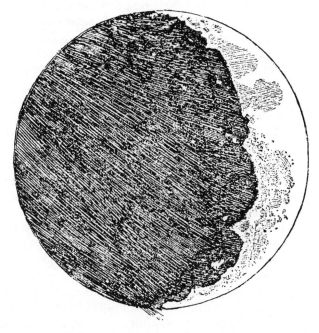

Let us note particularly that the said small spots agree always in having their darker parts directed toward the sun, while on the other side, opposite to the sun, they are crowned with bright contours, like shining summits. A similar sight occurs on earth about sunrise, when we behold the valleys not yet flooded with light although the mountains surrounding them are already ablaze with glowing splendor on the side opposite the sun. And just as the shadows in hollows on earth diminish as the sun rises higher, so those spots on the moon lose their blackness as the region illuminated grows larger and larger.

[64]

Again, not only is the boundary between light and shade on the moon seen to be uneven and wavy, but, still more astonishingly, many bright points appear within the darkened portion of the moon [though] completely divided and separated from the illuminated part and at a considerable distance from it. After a time, they increase in size and brightness, and within an hour or two become joined with the rest of the lighted part, which has meanwhile grown in size. At the same time more and more peaks shoot up as if sprouting, now here and now there, lighting up within the shaded portion; these become larger, and finally they too are united with that same luminous surface that extends ever farther. Examples of these are to be seen in the preceding figure. And on earth, before sunrise, are not the highest mountain peaks illuminated by the sun's rays while plains remain in shadow? And when the sun is fully risen, does not the illumination of plains and peaks become finally united? On the moon, the variety of elevations and depressions appears in every way to surpass the roughness of the earth's surface, however, as we shall demonstrate later.

And now I cannot pass over in silence something worthy of consideration that I observed when the moon was approaching first quarter, as shown in the preceding figure. Into the luminous part there extended a great dark gulf near the lower cusp. When I had observed this for some time, seeing it completely dark, a bright peak began to emerge after about two hours, a little below its center. Gradually growing, this presented itself in a triangular shape but remained completely detached and separated from the lighted area. Around it, three other small points soon began to shine; and finally, when the moon was about to set, this triangular form (which had meanwhile become more widely extended) joined with the rest of the illuminated region that suddenly burst into the shadowy gulf like a vast promontory of light, still encompassed by the three bright peaks already mentioned. Beyond the ends of the cusps, both upper and lower, certain bright points emerged that were well detached from the rest of the lighted part, as may also be seen depicted in the same figure.

There were also a great number of dark spots within the two horns [of light], especially the lower one, and those nearest to the boundary between light and shade appeared larger and darker, while those farther from the boundary [within the lighted horn] were not so dark and distinct. But in all cases, as was mentioned earlier, the blackish part of each spot is turned toward the source of sunshine while a bright rim surrounds the spot on the side away from the sun, in the direction of the shadowy region of the moon. This [lower] part of the moon's surface, where it is spotted as the peacock's tail is decked with azure eyes, resembles those glass vases that have been plunged while still hot into cold water and have thus acquired that crackled and uneven surface from which they receive the common name of *ice-cups*.

As to the large [and familiar] lunar spots, those are not seen to be broken in the above manner, full of cavities and prominences; rather, they are even and uniform, and brighter patches crop up in them only here and there. Hence, if anyone should wish to revive the old Pythagorean opinion that the moon is like another earth, its brighter parts might very fitly represent the land surface, and its darker regions that of the water. I have never doubted that if our globe were seen from afar, flooded with sunlight, the land areas would appear brighter and the watery regions darker.

Sagredo   I remember with amusement that when I first read this I thought the printer might have reversed what was written. But a little later

D 97–98   our friend, here in this courtyard, showed me how water poured on the bricks made them look darker, by rendering the surface smooth and altering the reflection of light from it. When I said that smoothness should brighten the reflection to approach that of a

D 71–83   mirror, he did not argue with me but simply hung a mirror on that wall, struck by the sun, and showed me that from most positions it appeared darker than the brick wall itself, rough though that was. That is what awakened my interest in optics.

Sarpi   You were fortunate both to have been shown the effect and to possess a mind curious about realities rather than overconfident of preconceived opinions grounded in reasoning badly applied. I know many who simply said that this book was filled with absurdities and looked no further into such of our friend's statements as appeared to them unreasonable.

Salviati   That was the usual response of philosophers, who were thereby

deprived of the joy of learning from this book new things on earth
and in heaven of which they did not dream. New knowledge is
reserved for people who do not already know everything, and do
not believe everything they have been told on ancient authority.
But back to the moon:

☆   The large spots in the moon are also seen to be less elevated than the
brighter tracts, for whether the moon is waxing or waning there are
always seen, here and there along the boundary between light and
shadow, certain ridges of brighter hue around the large spots, to
which we have attended in preparing the diagrams. The edges of
those [familiar] spots are not only lower, but also more uniform
[than the adjoining bright areas], being uninterrupted by peaks or
ruggedness.

Near the large spots the bright part stands out particularly, in such
a way that before first quarter, and approaching last quarter, in the
vicinity of a certain [familiar] spot in the upper, or northerly, region of
the moon vast prominences stand out, both above and below it, as
seen in the figures reproduced here.[13]

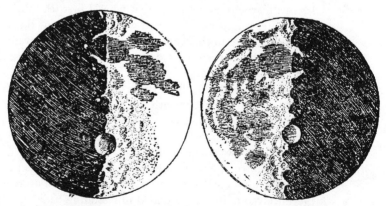

Before last quarter this spot is seen to be walled in [beyond the
lighted portion] by certain blacker contours that, like lofty mountains [still in shadow] appear darker on the side away from the sun and

lighter on that which faces the sun. That is the opposite of what happens in [small] cavities, where the part away from the sun appears brilliant while that turned toward the sun is dark and shadowy. After a time, when the lighted part of the moon's surface is diminished in size and the [large] spot is all (or nearly all) covered with shadow, bright mountain ridges gradually emerge from its shade. This two-fold aspect of the same spot is illustrated in the [last and] next figures.[14]

[67]

[68]

There is another thing I must not omit, for I beheld it not without a certain wonder, and this is that almost in the center of the moon there is a cavity larger than all the others and perfectly round in shape.[15]

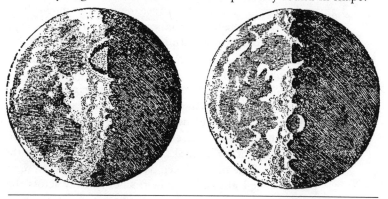

Sarpi   Excuse my interruption, but our friend's last diagram shows a cavity that is by no means larger than the ancient spots, which he has told us are lower than their boundaries and would therefore be called cavities. Moreover, though what he here depicted is perfectly circular, it is not shown at the center of the moon, but well below that—and still lower in a previous figure, where the illuminated portion of the moon is reversed. Can you explain these seeming discrepancies?

Salviati   Yes. The word *cavity* was reserved here for some appearances newly observed with the aid of the telescope, the ancient spots being called *depressions* rather than cavities. This circular cavity is "in the center" only in being equidistant from the moon's edges east and west, not also north and south. In fact, as you noticed, it is not always seen in the same place southerly from the moon's center, for the moon does not always present to us exactly the same face, as our friend will explain at some later time. I may add that the cavity in question is not as large as depicted in relation to the whole diameter, the drawings being intended to show the difference in its illumination when the sun is on one side of the moon or the other. Although vividly seen with the telescope, the effects could not be shown clearly in a small woodcut without giving the cavity a width greater than its true one. The figures drawn for this book were intended not as exact maps of the moon, but rather as illustrations of particular appearances described and explained in the text. To continue with that:

D 66

☆ I have observed this near both first and last quarters and have tried to represent it as correctly as possible in the diagrams above. As to light and shade, it offers the same appearance as would a region like Bohemia if that were enclosed on all sides by very lofty mountains disposed exactly in a circle. Indeed, this area on the moon is surrounded by such high peaks that its bounding edge on the side of the dark part of the moon is seen bathed in sunlight before the boundary between the light and shadow reaches halfway across the same area. As in the other spots, its shaded portion faces the sun while its lighted part is toward the dark side of the moon; and for the third time I draw attention to this as a very cogent proof of that ruggedness and unevenness which pervades the whole bright region of the moon. Among these spots, moreover, those are always darkest that touch the boundary line between light and shadow, while those farther from that appear both smaller and less dark, so that when the moon finally becomes full, at its opposition to the sun, the shadiness of the cavities is distinguished from the light of places [standing out] in relief by a lesser and rather meager difference.

The things we have reviewed are seen in the brighter regions of the moon. No such contrast of depressions and prominences is perceived

[69]

in the large spots as that which we are compelled to recognize in the brighter parts by changes of appearance that occur under varying illuminations by the sun's rays throughout the multitude of positions from which they strike the moon. In the large spots there are some areas rather darker than the rest, as we have shown in the illustrations. Yet these always present the same appearance, their darkness being neither intensified nor reduced—although, with some minute difference, they sometimes appear a little more shaded, and sometimes a little lighter, according as the sun's rays fall upon them more, or less, obliquely. Moreover, they join with the neighboring regions of the {large} spots in a gentle linkage whose boundaries mix and mingle.

It is quite different with spots that occur in the brighter region of the moon; those, like precipitous crags having rough and jagged peaks, stand starkly out in sharp contrasts of light and shade. And, inside the large spots, there are observed certain other, brighter zones, some of them very bright indeed. Yet both those and the darker parts present always the same appearance without change either of shape or of light and shadow, so that one may affirm without doubt that these [unchanging appearances] owe their existence to some real dissimilarity of parts. They cannot be attributed merely to irregular shape in which shadows move by reason of varying illumination from the sun, as is indeed the case with those other, smaller spots that occupy the brighter region of the moon and that change, grow, shrink, or disappear from one day to the next, owing their existence only to shadows cast by prominences.

But here I perceive that many persons will be assailed by uncertainty and drawn into a serious difficulty, feeling constrained to doubt a conclusion already explained and confirmed by many appearances. If that part of the lunar surface that reflects the sunlight more brightly is full of chasms—that is, countless prominences and hollows—why is it that the western edge of the waxing moon, the eastern edge of the waning moon, and the entire periphery of the full moon are not seen to be uneven, rough, and wavy? The edges look, on the contrary, as precisely round as if they had been drawn with a compass; yet the whole periphery consists of that brighter lunar surface that we have declared to be filled with heights and chasms. In fact, not a single one of the large [and smoother] spots extends to the extreme circumference of the moon, but all are collected together at some distance from the edge. Let me now explain the twofold reason for this troublesome fact, and in turn give a double solution to the difficulty.

In the first place, if the protuberances and cavities in the lunar body existed only along the extreme edge of the circular periphery bounding the visible hemisphere, the moon might (indeed, would necessarily) look to us rather like a toothed wheel terminated by a warty or wavy edge. Imagine, however, that there is not one single series of prominences [and depressions], situated just along the very circumference, but [that there exist] very many ranges of mountains, together with their valleys and canyons, disposed in ranks near the moon's edge—and not only in the hemisphere visible to us, but everywhere near the boundary line between the two hemispheres. Then an eye viewing them from afar would not be able to detect the separation of prominences by cavities, because the spaces between mountains situated around a given circle or [forming] a particular chain would be concealed by the interposition of yet other heights in yet other ranges. That will be especially true if the observer's eye is placed in the same straight line as the summits of the heights. It is thus that on earth, the summits of several mountains close together appear to be situated in one plane when the observer is a long way off and at an equal height. Likewise, in a rough sea the tops of waves will appear to lie in one plane, though between one high crest and the next there are gulfs and chasms of such depth as to hide not only the hulls, but even the bulwarks, masts, and rigging of stately ships. Now, since there are many chains of mountains and many chasms on the moon besides those at its periphery, and since the eye, regarding them from a great distance, lies nearly in the plane of the summits, no one need wonder that those appear as if arranged in a regular and unbroken line.

To the above explanation there may be added another; namely, that there exists around the moon's body, just as around the earth, a globe

[70]

of some substance denser than the rest of the aether. That may serve to receive and reflect the sun's rays without its being opaque enough to prevent our seeing through it, especially when it is not [itself] illuminated. Such a globe lighted by the sun's rays makes the body of the moon appear larger than it really is, and if thicker it could prevent our seeing the actual body of the moon. And it actually is thicker near the circumference of the moon—I mean, not in an absolute sense, but relatively to our visual rays, which there cut it obliquely. Thus it may obstruct our vision, especially when it is lighted, cloaking the lunar periphery that is exposed to the sun. This can be understood more clearly from the next figure, in which the body of the moon, ABC, is surrounded by the vaporous globe DEG:

[71]

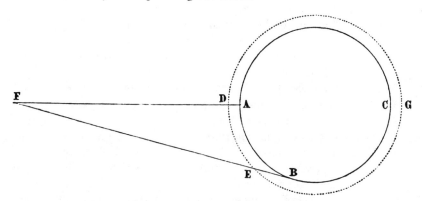

Our vision from F reaches the central region of the moon, at A for example, through a lesser thickness of the vapors DA, while toward the extreme edges a deeper vaporous stratum, EB, limits and shuts out our vision. One indication of this is that the illuminated portion of the moon appears larger in circumference than does the remainder of its orb which lies in shadow. And perhaps this same cause will appeal to some as reasonably explaining why the large spots on the moon are nowhere seen to reach its very edge, though it is probable that some should be present there. Possibly they are [made] invisible by concealment under a thicker and more luminous bulk of vapors.

Sarpi

This hypothesis bears an amusing resemblance to one made by our friend's adversaries against his assertion that the moon is not perfectly spherical and smooth, as philosophers have always maintained. I alluded to this earlier, and here seems a good place to speak of it. Doubtless Salviati knows it from the author himself, but Sagredo may not have heard of it.—You shake your head, so I go on. A Florentine philosopher named Lodovico delle Colombe sent his explanation to Rome in a letter to Father Clavius, after which Cardinal Joyeuse asked our friend for his reply. It was so clever that my correspondents at Rome, who are not overly fond of Jesuits, sent it to me.

xi, 118
xi, 141–55

Salviati

Indeed I do know of this. It happened that this same suggestion had been proposed earlier to a German astronomer whose other oppositions to lunar mountains had first been duly answered by our friend. Have you heard of it, Sagredo?

xi, 13

Sagredo

I am totally in the dark and wish to be enlightened.

Sarpi

Signor Colombe proposed to Father Clavius, who from the first had questioned the reality of lunar mountains and valleys, that observation be rendered impotent to prove the moon rough and irregular, as follows. Around the moon's body assume a shell of perfectly transparent crystal, through which our friend saw mountains and valleys formed in opaque material within. The crystalline surface of the moon remains nevertheless absolutely spherical and smooth, without a hair's deviation from perfection, as philosophers have always said.

xi, 118

Sagredo

Very clever, like so many creations by philosophers designed to save Aristotle from any fault whatever. I can imagine what our friend replied; tell me, and let me see if I have guessed aright.

Sarpi

He replied that he would grant to Colombe this absolutely transparent crystalline substance if that philosopher, with equal courtesy, would allow him then to form of it mountains ten times higher, and valleys as much deeper, than those he had observed and measured on the moon.

Sagredo   I like this better than what I guessed he would reply, because it is more vivid a way of showing that spherical shape did not follow from an imaginary material, but was dragged in from the very same Aristotelian assumption that Colombe proposed to save.—I guessed that our friend would merely reply that the suggestion was ingenious, but incapable of either proof or disproof.

Salviati   Do not reproach yourself, Sagredo, for in fact you guessed correctly. Our friend added that very statement in the same letter, after his more biting sarcasm that Fra Paolo's correspondents mentioned. He then went on to say that Colombe's proposal was like defining *the earth* as everything elemental lying between its center and a sphere passing through the summit of its highest mountain, and then saying that one had proved the earth to be perfectly round.

Sarpi   I am delighted to have this further information, Sagredo; too often our friend's witticisms are passed around without the serious thought that accompanied them. But while we are pausing to converse, I ask whether our friend still maintains his hypothesis of a lunar atmosphere like the earth's?

Salviati   No; he intends to retract that in another book. It was based solely on terrestrial analogy, as he believes all speculations about the heavenly bodies should be based. But the analogy now seems weak to him, because he has not been able to discover on the moon, by the most careful and repeated observations, any of the other things

D 100   associated with our atmosphere, such as clouds, haziness, or that wavering of images occasioned on earth by changing heat in the air. Since the lunar "day" and "night" are more than a dozen times the length of ours on earth, such changes should be very great in any lunar atmosphere. But let us continue with our friend's published conclusions:

☆   That the lighter surface of the moon is everywhere dotted with protuberances and gaps has, I think, been made sufficiently clear from the appearances already explained. It remains for me to speak of their dimensions and to show that the earth's irregularities [of surface] are

far less than the moon's—absolutely less, I mean, and not just relatively to the sizes of the two globes. This is plainly demonstrated as
follows.

I have often observed, in various situations of the moon with
respect to the sun, that some summits within the shadowy portion
appeared lighted, although they lay some distance from the boundary
of light [reflected to us from the sun.] By comparing that separation
to the whole diameter of the moon, I found that it sometimes ex-
ceeded one-twentieth of the diameter. Accordingly, let CAF be a great
circle of the lunar body, E its center, and CF the diameter, which is to
the diameter of the earth as two is to seven.

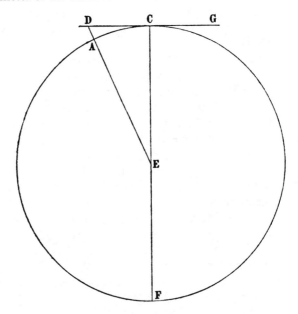

Since according to precise observations the earth's diameter is seven
thousand miles, CF will be two thousand, CE one thousand, and one-
twentieth of CF will be one hundred miles. Now let CF be the diameter of the great circle which divides the lighted part of the moon from

[72]   the dark part; for, because of the very great distance of the sun from the moon, this does not appreciably differ from a great circle [of the moon]. Let A be distant from C by one-twentieth of this [diameter]. Draw the radius EA that, when produced, cuts the tangent line GCD (representing the illuminating ray) at point D. Then arc CA, or rather the straight line CD, will consist of one hundred units whereof CE contains one thousand, and the sum of the squares of DC and CE will be 1,010,000. That equals the square of DE; hence ED will exceed 1,004, and AD will be more than four of those units of which CE contains one thousand. Therefore the altitude of AD on the moon, which represents a summit reaching up to the sun's ray GCD and situated at a distance CD from C, exceeds four miles. But on earth we have no mountains that reach a perpendicular height of even one mile. Hence it is quite clear that there are prominences on the moon loftier than those on earth.

Sagredo   Fra Paolo mentioned a while ago that a great furor was raised by Jesuits at Mantua about these lunar mountains, saying also that a German had attacked our friend's book on the same score. You, Salviati, have since spoken of a German whose oppositions were answered by our friend; was this the same man?

Salviati   It was; his name was Johann Brengger, and he was a friend of that Mark Welser of Augsburg to whom our friend addressed his *Sunspot Letters*.

Sagredo   It seems to me that this raises a question of procedure that we ought to decide. Shall we discuss oppositions that our friend consumed time in replying to as we go along, or shall we postpone that to another session when we have finished reading the book?

Salviati   I think we should not have a firm rule, but should deal with some objections as we go along and leave others for a later time. In the present instance, I see that the text here moves abruptly to another subject, so that it would not break any thread of thought if we intervened. Some objections, such as that which denied the existence of newly seen stars as illusions or defects of the lenses, could only be brought in now by main force, so to speak, and had better be deferred. We shall have enough objections to occupy a later session, even without this craggy one, so let me put the book down and turn to that.

x, 460   Toward the end of 1610 our friend received a letter from Brengger, forwarded to him by Mark Welser for comment. The German critic pointed out that in the central parts of the moon, great unevenness of the dividing line between light and dark parts does not permit accurate measurements of the kind described above. Moreover, when only two or three hours elapse between the first illumination of a peak in the darkened part and its subsequent joining with the lighted part, the peak would have moved but 1½°, and its height would need to be only one-third of a mile to catch sunlight at the beginning. A mountain four miles high, he said, implied an angle of 5° and a time of eight hours or more, longer than our friend claimed to have observed this phenomenon of a lighted peak that later merged with the sunlit part of the moon.

Sagredo   One moment, please. How was that angle of 1½° calculated?

Salviati   Brengger reasoned that sunlight takes over 29 days to circle around the whole moon, moving through 360°, so he calculated the motion during two or three hours to be that part of 360 which the said time is of 29 days.

Sarpi   That seems reasonable. Then, unless Brengger made a mistake in simple arithmetic, our friend's estimate of four miles does indeed seem far too high.

Sagredo   Excuse me, Fra Paolo, but I do not think that that follows. Looking back at our friend's first diagram, showing the appearance only four or five hours after new moon, I see a very thin but very long crescent. A week later, at first quarter, the dividing line between light and dark bisects the moon, as in his second figure. During that week, then, the top and bottom of the thin horns moved through a much smaller angle than did the center of the crescent, which during the same week traveled perhaps 75° or more. So it seems to me that we cannot just divide two or three hours by 29

days and take that fraction of 360° as if the resulting angle represented the advance of the bright edge during two or three hours,
no matter where on the moon the illuminated peak was seen. The
whole affair looks very complicated to me, since I can see how the
bright edge might advance several times as far, in the same time,
at the center than at the tip of a horn. That is all anyone would
need to say in reply to Brengger's argument. If our friend was
correct in estimating separation of a peak from the lighted part as
one-tenth of the moon's radius, then his calculation of the height
was properly made.

Sarpi   I see your point. But our friend's diagram shows a measurement at
the very rim of the moon, not near its center.

Sagredo   To illustrate the geometry of his measurement, he had to draw it
there. In fact, however, there are no isolated peaks in dark regions
near the rim of the moon, for reasons he has already set forth. And
since he here spoke of an illuminated peak in a large darkened
region, we must regard his geometrical diagram as turned through
a right angle and as showing the moon's axis, so to speak, pointed
toward our eyes, with the moon's equator shown as its rim. But
why should we argue, when we can just ask Salviati how our friend
replied?

Salviati   He wrote at once a very long answer, with many diagrams showing
x, 466–73   how circumstances alter the phenomena of illumination. Thus, a
long sloping mountain may seem to jump suddenly from an isolated point of light in a dark field to the outer boundary of the
region bathed in sunlight, whereas a steep, towering pinnacle
cannot do that. Again, some lesser peak, nearer to the lighted
boundary, could keep a more distant high mountain dark for some
time, so that only a short time might elapse from its first visibility
to its joining with the lighted region of the moon, though its peak
might stand miles above its base.

Sagredo   Such things did not occur to me, and even without them, the
whole matter looked complicated enough. I think you will agree,
Fra Paolo, that if our friend's way of analyzing the phenomena

| | exhibited by lighted peaks in dark regions fails to carry full |

exhibited by lighted peaks in dark regions fails to carry full conviction, no less does Brengger's supposed refutation of it. I suppose the German remained unsatisfied, Salviati?

Salviati Yes. He courteously urged his point again, and in the meantime Jesuits at Mantua began their battle and went to the absurd length of demanding no roughness whatever on the moon. There a learned Jesuit from Parma, after vainly urging moderation, presented our friend's analysis for consideration in purely geometric form. This, I believe, was Father Giovanni Blancani, an excellent mathematician, who nevertheless sided against our friend on other and more serious matters such as the discovery of sunspots. This question of lunar mountains was of concern mainly to philosophers who were poor mathematicians and went to absurd lengths to save the moon from any irregularity. One result was the imaginary crystal shell we mentioned earlier.

Sarpi Whose idea was that originally; do you know?

Salviati It was first suggested to Welser by some friend of his other than
xi, 13 Brengger, according to a letter he wrote to our friend early in 1611. Some say the idea originated with Father Christopher Scheiner, who soon became very angry at our friend over the issue of priority of sunspot observations. In Italy it was advanced by Colombe, who probably managed to think it up independently. He has a special
GAP 58–60 talent for saving Aristotle unharmed by any celestial phenomenon, as shown by his first book concerning the new star of 1604. But I think that enough has been said for now about the height of lunar mountains and the mathematical and philosophical oppositions they aroused. Our friend wrote a very long letter on this whole
xi, 178–203 matter to Father Christopher Grienberger at Rome, in September of 1611, and then dropped it and turned his attention to the study of more important things.

Sarpi So should we, in my opinion. If further thoughts about the lunar mountains occur to us, we can discuss them when we come to consider other objections against our friend's science. Please go on reading, Salviati.

Salviati At this point, as I said, the book takes up a new topic. Since I am a bit fatigued by my long excursion to the moon, I should be grateful if you would take up the reading.

Sarpi   With pleasure; give me the book. Our friend next writes:

☆   Here I wish to assign the cause of another lunar phenomenon well deserving of notice. I observed this not just recently, but many years ago, pointing it out to some of my friends and pupils, explaining it to them, and giving its true cause. Since it is rendered even more evident (and is easier to observe) with the aid of the telescope, I think it not unsuitable for introduction at this place, especially as it shows more clearly the connection between the moon and the earth.

When the moon is not far from the sun [in our line of sight] just before or after new moon, its globe not only offers itself to view on the side which is adorned with shining horns, but also a faint light is seen to mark out the periphery of the dark part that faces away from the sun, [that secondary light] separating this from the darker background of the aether. Now, if we examine this matter more closely, we shall see that not only does the extreme limb of the shaded side glow with this uncertain light, but the entire face of the moon (including the side that does not receive the glare of the sun) is whitened by a not inconsiderable glow. At first glance only a thin luminous circumference is noticed, contrasting with the darker sky coterminous with it, and the rest of the [moon's] surface appears darker by its touching the shining horns that distract our vision. But if we place ourselves so as to interpose a roof or chimney or some other object at a considerable distance from the eye, those shining horns may be hidden while the

[73]   rest of the lunar globe remains exposed to view. It is then found that that region of the moon, though deprived of sunlight, likewise shines not a little. The effect is heightened if the gloom of night has already deepened by departure of the sun, since in a darker field a given light appears brighter.

It is moreover found that this secondary light (so to speak) of the moon is greater according as the moon is closer to the sun. It diminishes more and more as the moon recedes from the sun, until it is seen very weakly after the first quarter and before the last, and uncertainly even when observed in the darkest sky. But when the moon is within 60° of the sun it shines remarkably, even in twilight—so brightly, indeed, that with the aid of a good telescope one may distinguish its large spots. This remarkable gleam has afforded no small perplexity to philosophers, some of whom have offered one idea and some another in order to assign it a cause. Some would say that it is an inherent and natural light of the moon's own; others, that it is imparted by Venus; still others, by all the stars combined; and others yet derive it from the sun, whose rays they would require to permeate the thick solidity of the moon.

Statements of that sort are refuted and their falsity evinced with little difficulty. For if this kind of light were the moon's own or were contributed by the stars, the moon would retain it and would display it particularly during eclipses, when it is alone in an unusually dark sky. Experience contradicts this, for the brightness seen on the moon during eclipses is much fainter and is ruddy, almost copper-colored, while this [secondary light] is brighter and is whitish. Moreover that other light is variable and movable, for it covers the moon's face in such a way that the place near the edge of the earth's shadow is always seen to be brighter than the rest of the moon. This undoubtedly results from touching by the sun's rays of some denser zone that surrounds the moon, by which contact a sort of twilight is diffused over the neighboring regions of the moon—just as on earth a sort of crepuscular light is spread in the morning and evening. But I shall *D 90* deal with this more fully in my book on the system of the world.

I interrupt my reading to point out that here we have the first mention of our friend's promised book, phrased as though it were already partly composed. I note that this comes in connection with a mere optical phenomenon, not regarded by most philosophers as even relevant to a book about so sublime and grave a subject as the system of the universe, and hardly worthy of consideration except by mere mathematicians. Occasional and incidental appearances are not taken by philosophers to be proper bases from which one may reason about the structure of the world; rather, they reason from certain and unquestionable metaphysical principles such as that which makes all heavenly bodies perfectly spherical and smooth. Now, our friend has mentioned that principle not to deduce anything from it, but only to reject it on observational grounds without offering any new principle in its place. Reasoning from earthly analogies is hardly an unquestionable certainty. From this first mention, therefore, it seems that he must have had in mind a book very different from any other yet published about the system of the world.

Sagredo One thing seems to me certain, and that is that our friend would object to the adjective *mere* before the word *mathematician*. In my opinion if *mere* has a proper place, it is before the word *philosophers*. Our friend, however, was scandalized when I said as much in a

xi, 379 letter to him.

Sarpi Verbose deductions by philosophers indeed become tiresome at times, but I hardly think our friend would wish to replace them entirely with the finicky precision of mathematicians. Have you something to say on this, Salviati?

Salviati I could say something at this stage, though I think we should leave our main discussions until later—at least until we have finished rereading this book, as I suggested at the beginning. But to offer you a hint at this point, I shall say that our friend would not use the adjective *mere* in speaking of serious mathematicians or philosophers, or indeed any group of men who soberly seek the truth. Some triflers give themselves such honorable titles, as Colombe

D 113 calls himself a philosopher; but our friend calls such men philosophasters, or memory experts who parrot Aristotle but do not philosophize. Above all he would not apply the word *mere* to any fact of sensate experience. Who knows that a seemingly trifling optical phenomenon cannot overthrow a whole train of deductions, mathematical or philosophical? The *Starry Messenger* is a book of "great, faraway, and admirable sights." Those are sensate experiences, and when our friend composes a book on the system of the world he will not ignore anything observed, however accidental or occasional, that might be brought against any of his conclusions.

D 55

In this he sides with Aristotle, who taught that a single contrary experience outweighs any amount of subtle reasoning.

Sagredo

CES 10–11

D 53–54

When Simplicio was here, Fra Paolo, Salviati told us that our friend now restricts his science to things evidenced by sensate experiences and necessary demonstrations, leaving anything that requires other evidence to philosophy, theology, and the humanities. Science he confines to a domain, however narrow, in which one should be able to avoid any gratuitous error.

Sarpi

Thank you; your hints content me to follow Salviati's advice and defer further discussions at least until we find some new clue to the contents of the book our friend had—and I hope still has—in mind. He was discussing, you remember, the faint light of the moon in its darkened part when it is thinly crescent. He continues thus:

☆

[74]

To assert that the moon's secondary light is imparted by Venus is so childish as to deserve no reply. Who is so ignorant as not to understand that from new moon to a separation of 60° between moon and sun, no part of the moon which is averted from the sun can possibly be seen from Venus? And it is likewise unthinkable that this light should depend upon the sun's rays penetrating the moon's thick solid bulk; for then this light would never dwindle, inasmuch as one hemisphere of the moon is always illuminated except during eclipses. And the light does diminish as the moon approaches first quarter, becoming completely obscured when that is passed.

Now since the secondary light does not inhere in the moon, and is not received from any star or from the sun; and since in the whole universe no body is left [to consider] except the earth, what must we conclude? What can be proposed? Surely we must say that the lunar body (or other dark and sunless orb) is [faintly] illuminated by the earth. Yet what is so remarkable about that? The earth, in fair and grateful exchange, pays back to the moon an illumination similar to what it receives from her throughout nearly all the darkest gloom of night.

Let us explain this matter more fully. At conjunction, the moon occupies a position between sun and earth; it is then illuminated by the sun's rays on the side that is turned away from the earth. The other hemisphere that faces the earth is cloaked in darkness, so the moon does not illuminate the earth's surface at all. Next, gradually departing from the sun, the moon comes to be lighted, partly, on the side turned toward us, and its whitish horns, still very thin, illuminate the earth with a faint light. Solar illumination of the moon then increases as the moon approaches first quarter, so a reflection of that light to the earth also increases. Soon the splendor on the moon extends to a semicircle and our nights grow brighter; at length the entire visible face of the moon is irradiated by the sun's resplendent rays and at full moon the earth's whole surface shines in a flood of moonlight. Now the moon, waning, sends us her beams more weakly and the earth is less strongly lighted; finally the moon returns to conjunction with the sun and black night covers the earth.

During this monthly period, then, moonlight gives us alternations of brighter and fainter illumination, a benefit repaid by the earth in equal measure. For while the moon is between us and the sun (at new moon), it faces the whole surface of that terrestrial hemisphere exposed to the sun and illuminated by vivid rays. The moon receives the light so reflected, and thus the nearer hemisphere of the moon—that is, the one deprived of sunlight—appears by virtue of this illumination to be not a little luminous. When the moon is 90° from the sun, it sees but half the earth illuminated (the western half), for the other (the eastern half) is enveloped in night. Hence the moon is itself illuminated less brightly from the earth, as a result of which the secondary light appears fainter to us. When in opposition to the sun, the moon faces the hemisphere of earth that is steeped in the gloom of night; if this position occurs in the plane of the ecliptic the moon will receive no light at all, being deprived of both the solar and the terrestrial rays. In its various other positions with respect to earth and sun, the moon receives more light or less according as it faces a greater or smaller portion of the earth's illuminated hemisphere. And between the two globes a relationship is maintained such that whenever the [nocturnal] earth is most brightly lighted by the moon, the moon is least brightly lighted by the earth, and vice versa.

Let these few remarks suffice us here concerning this matter, which will be more fully treated in our "System of the World." In that book, by a multitude of reasonings and valid experiences, solar reflection from the earth will be explained for those who declare the earth to be excluded from the circling of the stars by this most potent [argument] that it is devoid of motion and of light; [yet] wandering it is, and surpassing the moon in splendor—not the sink of all dull refuse of the universe—as we shall confirm by a wealth of physical and rational demonstrations.

D 63–67

Salviati   This second clue to the promised book, Fra Paolo, differs from the first. Though the topic being discussed is again a mere accidental and occasional optical phenomenon, as before, the question of the earth's motion is now suddenly brought in. No philosopher would deny that topic a place in a book on the system of the universe, though nearly all would deny that the earth has any motion. It is this question that our friend considers the most important of all to be dealt with in his future book, though I believe he did not mention the earth's motion again in the *Starry Messenger*. In fact, I had forgotten even this one mention of it, which I now see occurred here only incidentally, and because philosophers customarily link light and motion with heavenly bodies alone.

Sarpi   I was about to go on reading in accordance with my recent promise to defer such discussions, but since you provide me with an opportunity, Salviati, I am going to say something further. Our friend's theory about the primary cause of the tides, which we are going to discuss when I bring my notes of nearly twenty years ago, linked great motions of the seas with the Copernican motions of the earth. But since he has not spoken of that theory at Florence, it seems unlikely that he intends to include it in his book. Perhaps some serious flaw was later detected, and he abandoned it.

Salviati   Or, as I think more probable, some difficulty exists that he has not yet succeeded in removing, and he does not wish to discuss it until he is completely satisfied. Of course I am merely guessing, since I do not yet know the explanation of tides that he offered here long ago. What is important to me now is that what you just read implies an intention of including it, for he refers here to *physical* demonstrations that the earth both moves and also shines more brightly than the moon. Now, it is easy to think of physical demonstrations that the earth does reflect light brightly, but I can think of none that the earth moves, unless the tides offer physical demonstrations of that. I had not considered that possibility before and am more than ever eager to hear what he said to you. In short, our friend may have long preferred the Copernican system on grounds merely speculative rather than persuasive and may only now be considering difficulties to be resolved before putting physical grounds in print.

Sagredo   I shall say something of his early preferences when we hear Fra Paolo's notes. Even in this book our friend was cautious not to express full Copernican conviction, showing confidence only in the *Sunspot Letters* published this year. Do you know what changed his preference to conviction?

Salviati   It was several months after writing this book that our friend was first able to observe the moonlike phases of Venus, which can in no way be accommodated to Aristotelian cosmology or Ptolemaic astronomy, as they can be to the Copernican. Copernicus likewise failed to explain the relatively small observable variation in the brightness of Venus, considering the enormous changes in its distance from the earth as it circled the sun according to his new astronomy. That puzzle also arose in the old astronomy, in a different way. It was only after our friend saw with his own eyes that the explanation is found in the differing areas of Venus lighted by the sun as it circles that body that he began to say positively in letters to friends that Kepler and other Copernicans had philosophized correctly. Even after that, in the book he published earlier this year, he remained content to say only that all signs now point to

D&O 144   the ultimate victory of the Copernican astronomy.

Sagredo   That seems an excess of caution. Why not declare publicly that it was already victorious?

Salviati   Reflecting further on the phases of Venus, our friend realized that those changes would equally follow in the system of Tycho Brahe, in which all planets circle the sun while the sun carries them together around a stationary earth. It was therefore still necessary to remove all objections against a moving earth, to which Tycho himself had added a few of a physical character unknown to the

D 126–27   ancients, such as the equal range of a cannon from east or west.

Sarpi   Clearly our friend's proposed book will be much more than astronomical. Even more clearly, we shall never get to its present status at this rate; I therefore resume the reading:

    Thus far we have spoken of our observations concerning the body of the moon. Let us now set forth briefly what has been thus far observed regarding the fixed stars. And first of all, the following fact deserves consideration: stars, whether fixed or wandering, are not seen enlarged by the telescope in the same ratio as that with which it magnifies other objects, and even the moon. In the stars, this enlargement seems to be so much less that a telescope strong enough to magnify other objects one hundredfold [in area] is scarcely able to enlarge stars four or five times.[16] The reason for this is as follows.

    When stars are viewed by means of unaided natural vision, they present themselves to us not as of their simple (and so to speak their physical) size, but irradiated with a certain fulgor and fringed with sparkling rays, especially when the [darkness of] night is far advanced. From this they look to us larger than they would if stripped of those adventitious hairs of light, because the angle [formed] at the eye is determined not by the primary body of the star but by that brightness which extends widely around it. This is quite evident from the fact that when stars first emerge in twilight, they look very small at sunset even if they are of the first magnitude. Venus herself, when visible in broad daylight, is so small as scarcely to appear equal to a star of the sixth magnitude. Things fall out differently with other objects, including even the moon; whether seen in daylight or the deepest night, these appear always of the same bulk. Therefore stars are seen crowned among shadows, while daylight is able to remove their headgear—and not just daylight, but any thin cloud that interposes itself between a star and the observer's eye. A similar effect is produced by dark veils or by colored glasses, through interposition of which the stars are deserted by their surrounding brilliance. The telescope likewise accomplishes this same result. It removes from stars their adventitious and accidental rays and then it enlarges their simple globes (if indeed stars are naturally globular), so they seem to be magnified in a smaller ratio than are other objects [seen through the same telescope]. In fact, a star of the fifth or sixth magnitude when seen through a telescope presents itself as one of the first magnitude.

Also deserving of notice is the difference between the appearances of planets and fixed stars. Planets show their globes perfectly round and definitely bounded, looking like little moons, spherical and flooded with light all over; fixed stars are never to be seen bounded by a circular periphery, but have rather the aspect of blazes whose rays vibrate about them, and they scintillate a very great deal. Viewed with a telescope, they appear of a shape like that which they present to the naked eye, but enough enlarged so that a star of the fifth or sixth magnitude seems to rival the Dog Star, largest of all the fixed stars.

Now, beyond stars of the sixth magnitude, a host of other stars which escape the naked eye are perceived through the telescope, and these are so numerous as almost to defy belief. In fact, one may see more of these than all the stars contained in the first six magnitudes. The largest of them, which we may call stars of the seventh magnitude (or the first magnitude of invisible stars) appear through the telescope as larger and brighter than stars of the second magnitude seen with the naked eye.[17] In order to offer one or two proofs of their almost inconceivable number I adjoin pictures of two constellations; you may judge of the others from these as samples. In the first I intended to depict the entire constellation of Orion, but was overwhelmed by the vast multitude of stars and by limitation of time, so I have deferred that to another occasion. There are more than five hundred new stars distributed over one or two degrees of arc among the old ones. Thus,

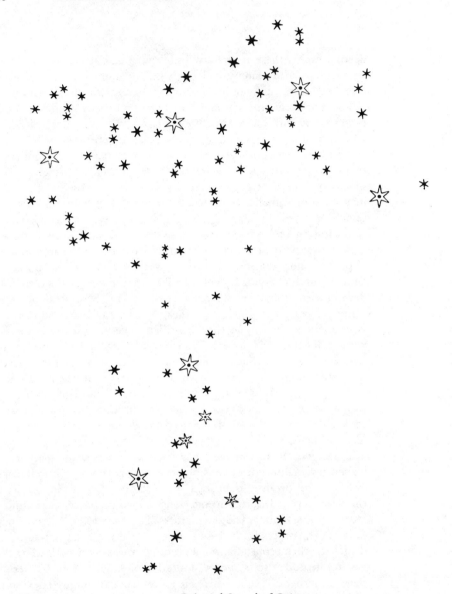

Belt and Sword of Orion

to the three stars in the Belt of Orion and the six in the Sword, previously known, I have added eighty adjacent stars recently discovered, preserving the distances between them as nearly as I was able. To distinguish the known (or ancient) stars I have shown them larger and outlined them doubly. The other (invisible) stars I have drawn small and without the added outline, also preserving the differences of magnitude as well as possible.

[78]   In the second example I have pictured the six stars in Taurus known as the Pleiades; I say six, because the seventh ["sister"] is rarely

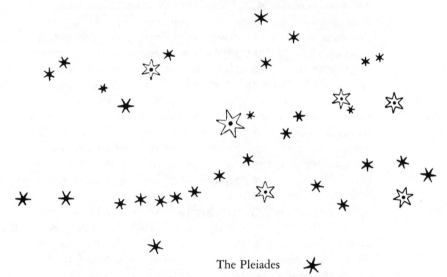

The Pleiades

visible. These lie within very narrow limits in the sky; near them are more than forty others, invisible, no one of which lies more than half a degree from the original six. Thirty-six of these are shown in the diagram; as in the case of Orion, I have preserved their intervals and magnitudes as well as the distinction between old stars and new.

Third, I have observed the nature and the material of the Milky Way. With the help of the telescope this has been scrutinized so directly and with such ocular certainty as to resolve all disputes that have vexed philosophers through so many ages.[18] The galaxy is in fact nothing but a congeries of countless stars, grouped in clusters. In every part of it to which the telescope is pointed, a great host of stars springs to view. Many of them are rather large and quite bright, while the number of smaller stars is incalculable.

But not only in the Milky Way are whitish clouds seen. Several patches of like aspect shine with faint light here and there throughout the aether, and when the telescope is turned toward any of them it confronts us with a tight mass of stars. Still more remarkable [than explanation of the Milky Way is the fact that] these stars, that have been called *nebulous* by every astronomer to the present, turn out to be groups of very small stars wonderfully arranged. Although each star separately escapes our sight because of its smallness or immense distance from us, the mingling of their rays gives rise to that gleam which was formerly believed to be some denser part of the aether capable of reflecting light from the stars or from the sun. Having observed some of these constellations, I decided to picture two of them. In the first you have the nebula called the head of Orion, in which I have counted twenty-one stars. The second contains the nebula called Praesepe [the Manger], which is not a single star but a mass of more than forty starlets, of which I have shown thirty-six in addition to the [two] Asselli, arranged as below.

We have now briefly recounted the observations made thus far with regard to the moon, fixed stars, and the Milky Way. There remains the matter which in my opinion deserves to be considered most important of all—disclosure of four PLANETS never seen from the creation of the world up to our time, telling the occasion of my having discovered and studied them, the arrangements of them, and observations made of their movements and changes during the past two months. I invite all astronomers to apply themselves to examine them and to determine their periodic times, something which has so far been quite impossible to complete owing to the shortness of time. Once more, however, I warn that it will be necessary to have a very accurate telescope such as we have described at the beginning of this discourse.

[79]

[80]

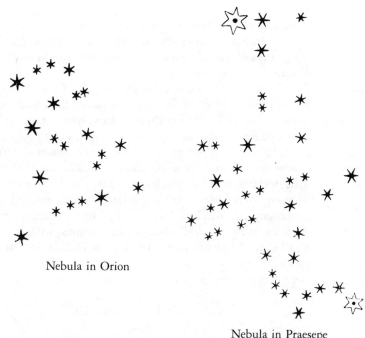

Nebula in Orion

Nebula in Praesepe

Sagredo	I see that we are now coming to that final discovery in which we said earlier that clues might be found to the perplexing problem of the powers of our friend's original telescopes. But we have already talked long today, and I think we should stop here, starting afresh tomorrow on the new topic just introduced.
Sarpi	Thank you for suggesting this, because I should return to my convent and arrange with Fra Fulgenzio Micanzio, my secretary and assistant, that he take over my duties for the next few days in order that I may be free to continue this interesting game of ours without interruption. I have been thinking also that if our plain monastic fare would be acceptable to you gentlemen, it would give great pleasure to me and to Fra Fulgenzio if you would sup with us

tonight. He especially, but also several other friars, used to join me and our friend in our discussions of natural motion and later of the disclosures of the telescope. I am sure they would like to hear news of him from you, Salviati.

Salviati   Our friend has spoken often of Fra Fulgenzio and I should be most happy to meet him. As for monastic fare, I hope you will not be offended, Sagredo, if I say that the sumptuous feasts you have provided for me during my stay in Venice, delightful as those have been, have indeed put me in the mood for a modest repast tonight.

Sagredo   I quite understand, for in my desire to make sure that you shall miss nothing of the delicacies of our city, both from our waters and from our commerce with foreign lands, I have reached a similar satiety myself. You have but to tell us when to arrive, Fra Paolo.

Sarpi   An hour or two should suffice for my business with Fra Fulgenzio. Until then, I bid you good evening and look forward to your coming.

Sagredo   Arrivederci, then; we shall not fail to be there.

End of the First Day

THE SECOND DAY

Sagredo	Good morning, Fra Paolo. Salviati and I have been recalling with pleasure our supper with you and Fra Fulgenzio, whom we look forward to seeing again this evening when we have finished our reading of our friend's book. We were just saying how surprising it was to find so many other friars also interested in hearing news of him and remarkably familiar with his work.
Sarpi	That is easily explained. When he was here, he often visited us and explained his discoveries in simple language, finding that to be more congenial than debating natural philosophy with professors who oppose his science. We Servites live more closely with ordinary people than do philosophers, so perhaps we share more in the general excitement over new wonders of nature. Several of the brothers used to join us in discussions, as happened last

night, and you could not have enjoyed your visit as much as they did.

Salviati   I fear that our reading today will seem less interesting than yesterday's, for the remainder of the book consists mainly in the recording of nightly positions of Jupiter's satellites, to use Kepler's term, or the Medicean stars as our friend christened them. Yet nothing could better serve to introduce certain difficulties that our friend faced in approaching his next discoveries, of which I have promised to give you an account later. As Sagredo takes up the reading, please remember that up to the time this book was published, particular satellites could not yet be distinguished by any visible mark and that daylight hours prevented uninterrupted recording of their motions. Only one starlet could be identified, and only sometimes, when it was separated from Jupiter by a large space into which the others never intruded. Long labor was required before the position of each satellite could be calculated for any given time. But that can now be done, so I shall be able to explain some of the puzzles that may arise as we read.

Sagredo   Then, since we may wish to refer to particular observations later, I shall assign a number to each as we go, to avoid the need of repeating the dates and hours of observation. To begin, then:

☆   On the seventh day of January in this present year 1610, at the first hour of night, when I was viewing the heavenly bodies with a telescope, Jupiter presented itself to me. And because I had prepared a very excellent instrument for myself, I perceived (as I had not done before on account of weakness of my previous instrument) that there were three starlets beside the planet, small indeed, but very bright.

Salviati   As agreed yesterday, let us begin our discussion with that question concerning the powers of our friend's telescopes. Further light is shed on this here, for he speaks not only of a newly prepared instrument but also of the inferior strength of one previously used in the observations before this memorable night.

Sagredo   We had reason to suppose that a telescope of 15 power was used in December 1609. This other, ready in the first week of 1610, must have been of at least 20 power on the basis of specifications we read yesterday for instruments capable of revealing all the things reported. But I do not believe we should assume that the newly prepared instrument was as powerful as the one shown later to Fra Paolo. It is not even necessarily true that the new starlets were completely undetectable with the weaker instrument, once they were known to exist. Our friend did not say here that they were, but only that he had not previously seen them by reason of weakness of the earlier telescope. I mention these things because it is possible, and even likely, that the observations we are about to consider were made with various telescopes. The new starlets might certainly leap suddenly into view if there was a doubling of power from 15 to 30. On the other hand it seems unlikely that an increase from 15 to 20 would alter complete invisibility into bright and clear visibility. The possibility may turn out to be of no consequence; however, let us keep it in mind and proceed with the reading:

☆ Though I believed them to belong to the host of fixed stars, they somewhat aroused my curiosity by their appearing to lie in an exact straight line parallel to the ecliptic, and by their being more splendid than other stars their size. Their arrangement with respect to Jupiter and to each other was as follows:

(1) East   *   * ◯   *   West

That is, there were two stars on the easterly side and one to the west; the more easterly star and the western one looked larger than the other. I paid no attention to the separations between them and from Jupiter, since at the outset I thought them to be fixed stars, as said before.

Sarpi Inasmuch as our friend had seen countless fixed stars previously unobservable, it is a wonder that he paid any attention to these at all.

Salviati   No, because it is only rarely that three fixed stars very near together lie in the same straight line, and in this case the line passed
also through the center of Jupiter and lay in the ecliptic, or very
near to that.

Sagredo   Yet those facts, though they aroused our friend's curiosity, did not
seem to him sufficiently remarkable to inspire further investigation, for the text goes on:

☆   But returning to the same investigation on the eighth of January,
led by I know not what, I found a very different arrangement.

Sarpi   Since he believed the starlets to be fixed, it would be natural to
expect a different arrangement, Jupiter having moved somewhat in
its orbit along the ecliptic. Unless, of course, Jupiter happened at
the time to be at "station," as astronomers call it when a planet
turns from direct to retrograde motion, or vice versa. But that
happens only occasionally, and Jupiter is never seen stationary
for long.

Sagredo   According to the tables of planetary motion, Jupiter was supposed
to be moving westward among the fixed stars, as it occasionally
does by its proper motion, which according to Aristotle is contrary
to the daily motion of the stars. What our friend saw this time did
not accord with that, for as he says:

☆   The three starlets were now all to the west of Jupiter, closer together,
and at equal intervals from one another as shown below:

(2) East   ◯   *   *   ✳   West

At this time I did not yet turn my attention to the manner in which
the starlets had gathered together, but I did begin to concern myself
with the question how Jupiter could be east of all these stars when on
the previous night it had been west of two of them. I commenced to
wonder whether Jupiter might not be moving eastward at this time,
contrary to the computations of astronomers, and had got in front of

[81]

them by that motion. Hence it was with great interest that I awaited the next night; but I was disappointed in my hopes, for the sky was then everywhere covered by clouds.

On the tenth of January, however, the stars appeared in this position with respect to Jupiter:

(3) East   ✷   ✶   ◯   West

That is, there were but two of them, both easterly, the third (as I supposed) being hidden behind Jupiter. As at the beginning, they were in the same straight line with Jupiter and arranged exactly in the line of the zodiac. Noticing this, and knowing that there was no way in which such alterations could be attributed to Jupiter's motion [alone], yet being certain that these were still the same stars I had observed [before]—in fact, no other star was to be found along the line of the zodiac for a long distance on either side of Jupiter—my perplexity was now turned into amazement. Certain that the apparent changes belonged not to Jupiter but to the observed stars, I resolved to pursue this investigation with greater care and attention. And thus, on the eleventh of January, I saw the following disposition:

(4) East   ✷   ✦   ◯   West

There were two stars, both to the east, the central one being three times as far from Jupiter as from the star farther east. The last-named star was nearly double the size of the other, whereas on the night before they had appeared about equal.

I now decided beyond all doubt that there existed in the heavens three stars wandering about Jupiter as do Venus and Mercury about the sun, and this became plainer than daylight from observations on the occasions that followed.[19] Nor were there just three such stars [as I was soon to learn]; four planets complete their revolutions around Jupiter, and I shall give a description here of their alterations as observed more precisely later on.[20] Also I measured the distances between them by means of the telescope, using the method explained earlier. Moreover, I recorded the times of observations, especially when more than one was made on the same night; for the revolutions of these planets are so swiftly completed that it is usually possible to note even their hourly changes.

Salviati   These first records, up to here, are less trustworthy than others that follow, through lack of mention of the exact times and by reason of the inadequacy of our friend's first attempts to estimate separations between the satellites, or between them and the edge of Jupiter's disk. Near that edge, moreover, they are difficult to see because of the bright light from that planet in the dark field of the sky, and the relative faintness of the newly discovered stars. For the same reason those stars look smaller when they first emerge close to Jupiter's disk than they appear an hour or two later, more distant from that brightness. These, and some other things which affected our friend's records of early observations, only gradually came to be understood by him in the light of continued experience.

Sagredo   All these things can be amplified later on, so rather than stray from the text let us proceed with our reading:

☆   Thus on the twelfth of January at the first hour of night I saw the [two outer] stars arranged in this way:

(5) East 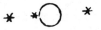 West

The more easterly star was larger than the westerly one, though both were easily seen and quite bright. Each lay about two minutes of arc distant from Jupiter.

[82]   (6) The third star [shown] was at first invisible and commenced to appear after two hours, when it almost touched Jupiter to the east and was extremely small. All were on the same straight line directed along the ecliptic.

Sarpi   Already I see one reason for Salviati's last comment, and here I want to note two things for his later attention. The first is that in speaking of "minutes of arc," our friend did not provide us—or astronomers—with any clear idea of what he saw, because he did not specify in any way how large the body of Jupiter itself may be. The second is that "almost touching Jupiter" must be inaccurate if, as you say, Jupiter's brightness would make it impossible to see a companion of Jupiter very close to its true edge.

Salviati   I shall reply to your first question now, deferring the other. In his first observations our friend used the apparent diameter of Jupiter when estimating the distances from its edge to a companion, or between two successive starlets. At the time of writing the *Starry Messenger* he took that diameter to represent about one minute of arc in the heavens, though last year he found, by his later measuring device, that it was less, usually about 40″.[21]

Sagredo   Thank you, for that suffices to give us a rough understanding of the positions here described. Later you must tell us how he could measure so tiny an arc in the sky. Next, he writes:

On the thirteenth of January four stars were seen by me for the first time, and in this situation relative to Jupiter:

(7) East   ✳ ◯✳ ✳ ✳   West

Three were westerly and one was to the east; they formed a straight line except that the middle westerly star departed slightly toward the north. The easterly star was two minutes of arc from Jupiter, and the intervals of the others from one another and from Jupiter were each about one minute. All the stars appeared to be of the same magnitude and, though small, were very bright—much brighter than are fixed stars of the same [apparent telescopic] size.

On the fourteenth, the weather was cloudy.

On the fifteenth, at the third hour of night, four stars were situated as follows with respect to Jupiter:

(8) East   ◯   ✳   ✳   ✳   ✳   West

All were westerly and very nearly arranged in a straight line; the one that was third from Jupiter was raised a bit to the north. That closest to Jupiter was smallest of all, the others appearing successively larger; the distances between Jupiter and the next three were equal and each of two minutes, while the westernmost was four minutes away from the one nearest to it. Though very bright they did not scintillate at all—and thus always they were seen, earlier as well as later. Now, at the seventh hour, only three stars were present, in this relation to Jupiter:

(9) East ⬤ ✦ ✦   ✦ West

They were exactly in the same straight line; the one nearest to Jupiter was quite small and was separated therefrom by three minutes; the second was one minute distant from this; and the third was four minutes thirty seconds from the second.

(10) After another hour the two middle stars were yet closer together; in fact, they were hardly thirty seconds apart.

Sarpi   There is some puzzle for me about the two observations depicted; that is, (8) and (9). The westernmost star shown in both must be the same one, and in fact this must be the outermost satellite, which moves most slowly of all. Yet in four hours it is reported to have moved 1½ minutes, or nearly as much as the starlet that was at first nearest to Jupiter and that had disappeared by moving still closer to it during the same four hours.

Salviati   Your perceptiveness, Fra Paolo, supplies an example appropriate to discussion of the question I deferred when you raised it. Our friend claimed no greater accuracy for his estimates of separation than one or two minutes, but you believe (and I agree with you) that he would not first estimate as ten minutes a separation from Jupiter that must have remained nearly the same for four hours, and then estimate it again as only 8½ minutes at the end of this time. The anomaly could indeed arise from individual errors of a minute or two in the successive separations that are added together for the westernmost, but some of those distances were themselves only one or two minutes, and we would not expect errors of the same magnitude as the estimates. We accordingly seek some other explanation, and it is Sagredo who has offered a possible clue to one.

Sagredo   Are you waiting for me to comment? I do not recall my having said anything except that more than one telescope may have been used in these observations.

Salviati   Yes; go on.

Sagredo   But that would not explain the seeming discrepancy between the two positions that Fra Paolo pointed out. Let the powers of two telescopes be what they may, our friend's counting unit of one Jovian diameter will differ in apparent size, but each will still remain the unit used in estimating separation. Surely separations are magnified in the same ratio as is the diameter of Jupiter, whatever the power of the telescope.

Salviati   Your reasoning is correct, Sagredo, but not an assumption that is concealed behind it. Let me offer a hint. Suppose that only one telescope was used for observations (8) and (9), but that on one occasion it was exactly focused and on the other it was not.

Sagredo   When a telescope is not exactly focused, an object seen through it appears indistinct and somewhat larger than it should. It has a somewhat fuzzy edge, so to speak. But surely our friend would detect that fuzziness.

Salviati   I believe he would, and would correct the focusing, though sharp focus is not as evident for a bright disk seen against a dark sky as it is for a church window observed in daytime. But I did not ask my question in order to suggest this kind of error, though I believe it might arise; my purpose was another one.

Sarpi   I think I see your point. Exact and inexact focusing of the same telescope would result in alteration of the counting unit without affecting the separations between satellites, but only the apparent separation from Jupiter of the nearest satellite on either side of the apparent edge of the bright disk. In the same way, it might happen that *exact* focusing of two *different* telescopes could alter the counting unit, quite apart from the alteration occasioned simply by difference of power of the two instruments.

Salviati   It might indeed. Behind your reasoning, Sagredo, lay the assumption that two telescopes could differ in power but not in some other quality that could affect the apparent size of Jupiter's disk.

Sagredo   Fra Paolo, say no longer that politics has dimmed your scientific faculties! Your hint is that if a lens is not perfectly ground to a spherical surface, the effect will be like that of inexact focusing. The telescope will then speciously enlarge Jupiter's disk, but will still correctly show a separation between two satellites that appear as mere points of light. So if two different telescopes were used, one—or for that matter both—having some imperfection of this kind, discrepancies would be expected between the counts made of apparent Jovian diameters between two satellites. It would be some time before our friend would become aware of this, and it is not easy to grind lenses to exact spherical surfaces. At the beginning he may have used lenses that gave Jupiter a spurious ring of light, so to speak, that would in turn extend the area within which it is hard to see a satellite very near to the bright disk.

Salviati   That was the point of my interrogations—not to assert that the discrepancy between observations (8) and (9) arose from differences in focusing the same telescope on two occasions, or from the use of telescopes differing in power and also in the size of that spurious ring of light around the bright disk, but to have you understand clearly some of the difficulties attending these earliest estimates of

x, 440   satellite positions. Even now, when our friend has installed machinery for the grinding of lenses at his house in Florence, there are still problems in making them exactly spherical.

Sarpi   What has been said answers my previous inquiry about a satellite described as "nearly touching" Jupiter. A spurious ring of light around Jupiter, occasioned either by imperfect focusing or by imperfection of lenses, would not only affect the unit used in counting satellite separations; it would also invade or extend the area of difficult visibility in which a lesser light near Jupiter cannot be seen. The planet itself nevertheless always appeared to our friend as a little round moon, distinctly bounded in the dark sky, by reason of the great contrast between Jupiter's bright light and darkness.

Salviati   I believe we should adopt that explanation, though the problems raised by special conditions of visibility close to Jupiter's disk are quite difficult to understand completely. For our present purpose it suffices that when our friend began his journal of positions of the Medicean stars, there were many impediments to sharp and clear observation, to say nothing of exact measurement. Some problems vanished with increasing familiarity by repeated observations, but precise measurements became possible only long after the book was published. Our friend first became aware of errors and confusions in the printed record of his observations when, toward the end of 1611, he hinted to his Roman friend and correspondent Signor Giovanbattista Agucchi how the periods of the companions of Jupiter could be approximately obtained by careful study of the *Starry Messenger* alone, if one only knew the maximum elongation for each of the starlets. He then apologized for an error in the book which at first induced Signor Agucchi to suppose an epicycle for the outermost satellite.

GAW
175–76

Sagredo   We expect to hear from you later how these lingering unclarities were finally conquered by our friend, so let us now return to our reading.

☆   On the sixteenth at the first hour of night I saw three stars in this order:

(11) East   West

Two had Jupiter between them and lay distant from it forty seconds on either side; the third, westerly, was eight minutes from Jupiter. The two nearest to Jupiter appeared not larger, but brighter, than the more distant one.

Sarpi   That seems remarkable, since ordinarily these stars are seen as faint when close to Jupiter's bright disk, growing larger as they retreat into the dark field of the sky.

Salviati   In fact there was a good clue here to the problem of identifying an individual companion of Jupiter on different nights. Our friend did not know at the time, as he does now, that the star moving in the orbit most remote from Jupiter is somewhat fainter than the others (except when others are faint merely by proximity to Jupiter's bright disk). Calculations from his later tables show the westerly star here to have been that fainter one, which he eventually came to designate by four dots and which we may conveniently speak of as satellite IV,[22] numbering them all by the respective sizes of their Jovian orbits.

Sagredo   To continue:

    On the seventeenth, thirty minutes after sunset, the configuration was thus:

(12) East West

A single easterly star was three minutes distant from Jupiter, and likewise there was only one to the west, eleven minutes from Jupiter. The easterly star appeared twice as large as the westerly, and there were but these two {at first}. But after four hours, about the fifth hour {after sunset}, a third star began to emerge on the east that I believe must earlier have been conjoined with the {easterly} one shown above, and this was the position:

(13) East   West

The middle star, as close as possible to the easternmost, was only twenty seconds distant from that and declined a little to the south from the line passing through the two outermost stars and Jupiter.

Sarpi   I wonder where the missing star hid for four hours and a half. Since the westernmost was IV, and the other two moved apart only a short way in all this time, I suppose that those were II and III. That would leave the most rapidly moving star of all as the one hidden for over four hours.

Sagredo   You are right on all counts. The missing starlet was indeed satellite

I, as determined by later calculation. At observation (12) it had just become imperceptible by moving into the region of impaired visibility near Jupiter's bright disk. At (13) it had moved directly behind the center of Jupiter. To continue with the text:

On the eighteenth, twenty minutes after sunset, this was the appearance:

(14) East   West

The eastern star, larger than the western, was eight minutes distant from Jupiter; the western one was ten minutes from Jupiter.

On the nineteenth, at two hours of night, this was the order of these stars:

(15) East   West

That is, there were three stars in a perfectly straight line with Jupiter. One, easterly, was distant six minutes from Jupiter; between Jupiter and the first following to the west ran an interval of five minutes; and this in turn was four minutes distant from the westernmost.

I next stood in doubt whether midway between the easterly star and Jupiter there was not a tiny star, so close to Jupiter as almost to touch it.[23]

(16) East   West

And indeed at the fifth hour I saw clearly that this {now} occupied exactly the midpoint between Jupiter and the {above} easterly star, so that this was the position:

(17) East   West

Moreover, the last star to appear was very small, though at the sixth (18) hour it was of a magnitude almost equal to the others.

On the twentieth, at one hour and a quarter [after sunset], an arrangement like this appeared:

(19) East   . ◯ .. .   West

There were three very small stars, barely to be seen, and the distances between them did not exceed one minute; I was uncertain whether there were two stars or three to the west. Around the sixth hour they were arranged as follows:

(20) East   * ◯ ..   West

The easterly star was twice as far from Jupiter as before; that is, [it was distant] two minutes; the middle westerly star was forty seconds from Jupiter, and the westernmost was twenty seconds farther. Finally, at the seventh hour, three little stars were seen to the west:

(21) East   . ◯ .. .   West

The nearest to Jupiter was twenty seconds from it; between that and the westernmost was an interval of forty seconds; between them there was seen another, deviating a bit southward and not more than ten seconds from the westernmost.

Sarpi   Could our friend really make such fine distinctions?

Salviati   At first he could not, but we are now seeing what could be done in the light of experience of a score of observations. Two things are to be noted, however. The first is that to distinguish the separation between two points of light in the dark field of the sky, far from the bright disk, and to judge how many such separations would make up one counting unit (that is, Jupiter's apparent diameter), is easier than it is even to see a starlet very close to Jupiter, let alone to judge its separation from the apparent edge when its own light seems very faint. The other is that because the counting unit has appreciable size it is not hard to judge its half, and its quarter, or to estimate when some distance is less than one-quarter of it—though beyond that, most of us would not care to attempt greater precision.

Sagredo Since our friend did not describe any separation as less than ten seconds, I presume that one-sixth of the apparent diameter of Jupiter was the smallest distance that he could discern between planet and starlet, when not merged into one light at small separations.[24]

Salviati Yes; there was a bare minimum separation below which two satellites distant from Jupiter were seen as one. To that minimum our friend applied the estimate of ten seconds.

Sagredo Very reasonable. Well, to go on:

☆   On the twenty-first, half an hour after sunset, there were three stars east of Jupiter, equally distant from one another and from Jupiter:

(22) East   ◂ * ◂ ◯   *   West

The intervals, as I estimated them, were of fifty seconds; there was also one star to the west, four minutes distant from Jupiter. The star nearest to Jupiter on the east was smallest of all, the other three being somewhat larger and about equal to one another.

On the twenty-second, at the second hour, the arrangement of stars was this:

(23) East   *   ◯ •,   *   West

[85]   From the easterly star to Jupiter there was an interval of five minutes, and from Jupiter to the most westerly, of seven minutes. The middle two on the west were forty seconds apart, the nearer to Jupiter being one minute from it. These two intermediate stars were smaller than the outer ones. All lay in the same straight line along the zodiac

except that the middle star of the westerly three deviated a bit south-ward. But at the sixth hour of night they appeared in the position:

(24) East   •   ⬭ • ✳ ✲   West

The easterly star was very small and was, as before, five minutes distant from Jupiter. The three to the west were equally distant among themselves and from Jupiter, each interval being about one minute twenty seconds. The star nearest Jupiter appeared smaller than the two next following, and all appeared arranged along the same straight line.

On the twenty-third, forty minutes after sunset, the location of the stars was about as follows:

(25) East   ✳   • ⬭   ✳   West

There were three stars, in a straight line along the zodiac as always; two were easterly and one westerly. The more easterly was seven minutes from the next, which was two minutes forty seconds from Jupiter, while Jupiter was three minutes twenty seconds from the westerly star. But at the fifth hour the two stars that had been at first closest to Jupiter were no longer to be seen, I believe because they were hidden by Jupiter, and the appearance was thus:

(26) East   ✳   ◯   West

Sarpi   This is very strange—two stars moving in opposite directions at very nearly the same speed, through considerable distances, yet without the fourth ever making an appearance during quite a long time.

Salviati   Do not forget that the body of Jupiter plus its adjoining ring of light and region of difficult visibility of these faint stars could account for as much as two minutes in the estimates made in this book. Hence the speeds of the two starlets near Jupiter need not have been nearly equal. And as we have seen before, even the most rapidly moving satellite may linger several hours conjoined with Jupiter, or too close to the bright disk for our friend to see it.

Sagredo   As you have said, these situations could be later investigated by means of our friend's tables of the motions, once he became able to distinguish and identify each star at any time.

Next, he writes:

☆   On the twenty-fourth three stars were seen, all to the east and almost in the same straight line with Jupiter:

(27) East ✹   ⬤   West

The middle one alone seemed to deviate slightly southward. The closest to Jupiter was two minutes from it, and the next was thirty seconds from this, while the easternmost was nine minutes beyond that. But at the sixth hour only two stars presented themselves, in this position:

(28) East ✹   ✹   ⬤   West

[86]   that is, exactly in a straight line with Jupiter, from which the closer was three minutes distant and the other was eight minutes from that. If I am not mistaken, the two intermediate starlets first observed had merged into one.

On the twenty-fifth, at one hour forty minutes [after sunset], such was the arrangement:

(29) East   ⬤   West

There were in fact only two stars, to the east and also quite large. The easternmost was five minutes from the middle one, which was six minutes from Jupiter.

On the twenty-sixth at forty minutes [after sunset] the order of the stars was this:

(30) East    West

That is, three stars were seen, of which two were to the east and the third to the west of Jupiter; the last-named was five minutes from it; the middle star east of Jupiter was five minutes twenty seconds away; and the easternmost was six minutes from the middle one. They were situated in the same straight line and were equal in magnitude. Later, at the fifth hour, the arrangement was much the same, differing only in that near Jupiter a fourth star emerged to the east, then distant thirty seconds [here the printer has *minutes*] and lifted a little to the north from the straight line [of the others], as shown in the next figure:

(31) East    West

On the twenty-seventh, one hour after sunset, only a single starlet was seen, to the east and in this position:

(32) East   ✴   ◯   West

It was very small and seven minutes distant from Jupiter.

It was impossible to make observations on the twenty-eighth and twenty-ninth by reason of interposed clouds.

On the thirtieth, at the first hour of night, the stars were seen situated in this way:

(33) East   ✴   ◯   ✴   West

One was easterly two minutes thirty seconds from Jupiter, and two westerly; the one closer to Jupiter was three minutes from it, and the other one minute from that. The places of the more distant stars and Jupiter were in a straight line, but the intermediate star was a little to the north. The westernmost was smaller than the others.

[87]   The last day [of January], at the second hour of night, two stars appeared east of Jupiter and one to the west:

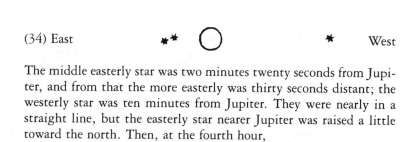

(34) East   West

The middle easterly star was two minutes twenty seconds from Jupiter, and from that the more easterly was thirty seconds distant; the westerly star was ten minutes from Jupiter. They were nearly in a straight line, but the easterly star nearer Jupiter was raised a little toward the north. Then, at the fourth hour,

(35) East   West

the two easterly stars were still closer together, as they were separated by only twenty seconds. In these observations the western star appeared rather small.

On the first day of February at the second hour of night, the position was thus:

(36) East   West

The more easterly star was six minutes distant from Jupiter; the westerly, eight minutes. On the eastern side a very small star was twenty seconds from Jupiter. They determined a perfectly straight line.

On the second of February the stars appeared in this order:

(37) East   West

To the east, one [star] alone was six minutes distant from Jupiter. Jupiter was four minutes from the nearer to the west, while from that to the westernmost was an interval of eight minutes. They lay exactly in the same straight line and were of about equal magnitude. But at the seventh hour there were four stars:

(38) East   West

Jupiter occupied the central place. Of these stars, the more easterly was four minutes from the next, which was one minute forty seconds from Jupiter; Jupiter was six minutes from the nearer westerly; this, eight minutes from the westernmost. They were all in the same straight line along the zodiac.

Sarpi   Here the distances in minutes imply excessive motion of satellite IV, which was obviously the westernmost on this night.

Salviati Fra Paolo, you have put your finger on a discrepancy that was destined to have a most interesting history. I do not wish to interrupt our reading to speak of that now, except to say that it was this night's figures that confused Signor Agucchi in the matter mentioned earlier, and for which our friend apologized to him. Had he deliberately wanted to baffle other astronomers, he could not have found a better means than that provided by a simple error in one number reported for this last observation. Even more important was a published attack against our friend's book based on this same inconsistency that you have noted, as I shall explain after we have finished our reading.

Sagredo   It was hardly to be expected that a first report of so many novel and difficult observations would be free from mistakes, whether by our friend or by the printer. Apparently this error occurred in such a way that it seriously misled more than one reader.[25] I can hardly wait to hear how that came about, so I hurry on now with my reading:

☆   On the third, at the seventh hour, the stars were arranged in this series:

(39) East   ✳ West

[88]   The easterly star was one minute thirty seconds from Jupiter, the nearer westerly was two minutes; from the latter, the more westerly star was ten minutes distant. They were precisely in the same straight line and were of equal magnitude.

On the fourth at the second hour four stars stood around Jupiter, two easterly and two westerly, arranged in a perfectly straight line as in the next figure:

(40) East  West

The easternmost was three minutes from the next, which was forty seconds from Jupiter; Jupiter was four minutes from the next westerly, which was six minutes from the westernmost. They were about equal in magnitude but the nearest to Jupiter appeared a bit smaller than the rest. Then, at the seventh hour, the easterly stars were only thirty seconds apart:

(41) East ✸✷ ◯ ✷ ✷ West

Jupiter was two minutes distant from the nearer easterly star and four minutes from the nearer westerly star, which was three minutes from the westernmost. They were all equal and on the same straight line along the ecliptic.

On the fifth the sky was cloudy.

On the sixth only two stars appeared, with Jupiter in the middle as seen in this figure:

(42) East ✷ ◯ ✷ West

The easterly star was two minutes from Jupiter; the westerly, three minutes; they were in the same straight line with Jupiter and were equal in magnitude.

On the seventh there were two stars, both east of Jupiter, disposed in this way:

(43) East ✷ ✷ ◯ West

The separations between them and Jupiter were equal, that is, one minute, and a straight line passed through them and Jupiter.

On the eighth at the first hour there were three stars, all easterly, as in this figure:

(44) East ✷✷ ✷ ◯ West

[89]

The nearest to Jupiter, quite small, was distant one minute twenty seconds from it; the middle one was four minutes from this, and the easternmost, very small, was twenty seconds from the middle one. I was dubious [at this time] whether closest to Jupiter there was only one star, or two; in fact, at moments it appeared that there was beside it, to the east, another one remarkably small and separated from it by only ten seconds. All were in the same straight line extending along the path of the zodiac. Now, at the third hour the star nearest Jupiter (45) [as shown above] almost touched Jupiter, being distant from it only ten seconds; meanwhile the others had become more distant from Jupiter, since the middle one [as shown above] was now six minutes from Jupiter. Finally, at the fourth hour, the one which had (46) been nearest to Jupiter [an hour before] conjoined with it and could no longer be distinguished.

Salviati	You see that by this time, Fra Paolo, our friend felt able to speak confidently of but ten seconds separation from Jupiter's disk.
Sarpi	Truly practice makes perfect in measuring by eye alone.
Sagredo	Or at least experience heightens confidence. Next, he writes:

On the ninth at thirty minutes after sunset there were two stars to the east and one to the west, in this position:

(47) East   ◯   West

The most easterly, which was rather small, was four minutes from the next; the middle one, larger, was seven minutes from Jupiter; and Jupiter was four minutes from the westerly [star], which was small.

On the tenth, one hour thirty minutes after sunset, two very small stars appeared, both to the east, in this position:

(48) East   ★   West

The more distant was ten minutes from Jupiter; the nearer, twenty seconds; and they were in the same straight line. But at the fourth

(49) hour the star nearer to Jupiter no longer appeared, and the other looked so diminished as hardly to be perceived, though the air was quite clear. It was farther from Jupiter, since it was twelve minutes distant.

On the eleventh at the first hour there were two stars to the east and one to the west:

(50) East  West

The westerly star was four minutes from Jupiter; the closer star on the east was likewise four minutes from Jupiter, and the easternmost was eight minutes from that. They were quite bright and in the same straight line. But at the third hour a fourth star appeared very close to Jupiter on the east, smaller than the others:

(51) East  West

It was separated from Jupiter by thirty seconds and was raised a bit northward from the line designated by the other stars. All were very splendid and easily visible. Now, at five hours thirty minutes, this eastern star nearest Jupiter, already more distant from it, occupied the place midway between Jupiter and the next easterly star:

[90]

(52) East  West

They were all in the same straight line and were equal in magnitude, as may be seen in the figure.

On the twelfth, forty minutes after sunset, two stars were seen to the east and likewise two to the west. The easternmost star was ten minutes distant from Jupiter, and the westernmost, eight minutes:

(53) East West

Both were quite visible; the other two, near Jupiter, were very small, especially the eastern one, which was forty seconds from Jupiter while the westerly one was one minute. But at the fourth hour the little star (54) that had been close to Jupiter to the east no longer appeared.

Salviati   Later calculations show that whatever had been recorded first as close to Jupiter on the east cannot have been one of the Medicean stars; perhaps a fixed star lay exactly in line with them.

Sagredo   Maybe this was one of those illusions created by the lenses to which some at first attributed all of our friend's new discoveries in the heavens. But, to continue:

☆

On the thirteenth, one half hour after sunset, two stars appeared to the east and two to the west:

(55) East 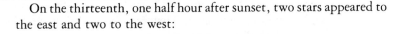 West

The easterly star closer to Jupiter, quite bright, was two minutes from it; from that the more easterly one, less visible, was four minutes distant. Of the westerly stars the one farther from Jupiter, easily visible, was distant from it four minutes. Between this and Jupiter was interposed a little starlet closer to the westernmost star, from which it was distant no more than thirty seconds. All were in the same straight line exactly along the ecliptic.

On the fifteenth (because the sky was covered by clouds on the fourteenth), at the first hour, this was the position of the stars:

(56) East  West

Thus there were three easterly stars while none was to be seen to the west. The nearest to Jupiter was fifty seconds distant from it; the next was twenty seconds from this; and from it the easternmost was then two minutes away and was larger than the others. The one nearest to Jupiter was in fact very small. About the fifth hour only one of the

(57) East  West

[91]

stars close to Jupiter was perceived, distant thirty seconds from Jupiter, while the easternmost had increased its distance from Jupiter and was now four minutes away. Then, at the sixth hour, in addition to the two already mentioned [as remaining visible] to the east, a very small star appeared to the west, two minutes from Jupiter.

(58) East   ✷   •◯ ✶   West

Sarpi   How could it first appear so far from Jupiter?
Salviati   That happens when there are eclipses, though calculation shows there was none this night. The emerging star was IV, always faint, but it should have been visible at the fifth hour.
Sagredo   Perhaps haze on the lens or in the sky concealed it, or a weak telescope may have been used. Next, we read:

☆   On the sixteenth, at the sixth hour, they stood in this position:

(59) East   ✷   ◯   ✷   ✷   West

That is, the easterly star was seven minutes distant from Jupiter, which stood five minutes from the next westerly, and that was then three minutes from the westernmost. They were all of about the same magnitude, easily visible, and in the same straight line exactly along the path of the zodiac.

On the seventeenth, at the first hour, there were two stars:

(60) East   ✷   ◯   ✶   West

One was easterly, three minutes distant from Jupiter, and the other westerly, distant ten minutes; this was somewhat smaller than the easterly star. At the sixth hour the easterly one was closer to Jupiter, (61) being fifty seconds from it, while the westerly one was farther away; that is, twelve minutes.[26] In both observations they were found in the same straight line and were both rather small, especially the easterly star at the second observation.

On the eighteenth at the first hour there were three stars, of which
two were westerly and one easterly; the latter was three minutes from
Jupiter. The nearer westerly star was two minutes distant [from
Jupiter] while the westernmost was eight minutes from the middle
one:

(62) East  West

All were exactly in the same straight line and were about of equal size.
But at the second hour the stars nearest Jupiter were equidistant from
(63) it, the westerly one being also three minutes away. Then, at the
sixth hour, a fourth little star appeared between the easterly star and
Jupiter in this arrangement:

(64) East  West

The more easterly was distant from the less by three minutes, and that
from Jupiter by one minute fifty seconds.

Salviati   There is a printer's error here; the text should read "one minute, or
fifty seconds."

Sagredo   I see what you mean, for he goes on to say:

☆   Jupiter was three minutes from the next westerly star, which was
seven minutes from the westernmost. They were about equal, except
that the easterly star near to Jupiter was a bit smaller than the rest,
and they were in the same straight line parallel to the ecliptic.

Sarpi   I do not see how that confirms the printer's error.

Sagredo   If I am not mistaken, the area near Jupiter in which the starlets
could not be easily seen extends about one minute from the apparent edge, or a little less, and there they appear to be smaller than
the others. At a distance of nearly two minutes, as the printer has
it, that effect would not be expected.

Salviati	That is right, and later calculation confirms the positions of both easterly stars after this correction.
Sagredo	I have added "or" in my copy; the text goes on:

☆ On the nineteenth, forty minutes after sunset, only two stars escorted Jupiter, both westerly:

(65) East   ◯   ✳   ✳ West

They were quite large and perfectly in line with Jupiter along the path of the ecliptic. That nearer to Jupiter was seven minutes distant from it, and six minutes from the westernmost.

On the twentieth the sky was cloudy.

On the twenty-first, an hour and a half after sunset, three quite small stars were seen in this arrangement:

(66) East   ✳   ◯   ✳   ✳   West

The easterly one was two minutes from Jupiter, which was three minutes from the next to the west, that was seven minutes from the westernmost; they were exactly in the same straight line parallel to the ecliptic.

On the twenty-fifth (since the three preceding nights were cloudy), at one hour thirty minutes [after sunset], three stars appeared:

(67) East   ✳   ✳   ◯   ✳   West

The distances between the two to the east and Jupiter were equal, being four minutes; the westerly one was two minutes from Jupiter. They were exactly in the same straight line along the path of the ecliptic.

On the twenty-sixth, half an hour after sunset, there were only two stars:

(68) East ✳   ◯   ✳   West

One was easterly, distant from Jupiter ten minutes, and the other westerly, distant six minutes. The easterly star was somewhat smaller than the westerly. At the fifth hour three stars were seen:

(69) East ✳   ⬤✳   ✳   West

Besides the two already mentioned, a third was perceived near Jupiter to the west, very small, that had at first been hidden under Jupiter and was not one minute distant from it. The easterly star appeared farther away than before, being eleven minutes from Jupiter.

Sarpi   Excuse me, Salviati, but does not *sub Iove*, "under Jupiter," mean that the newly appearing star had been previously behind Jupiter?

Salviati   No indeed, for it was clearly moving westward from Jupiter. It had been "under" in the sense of "lower than Jupiter," and closer to us. For the satellites all move in the same direction as Jupiter's proper motion, which is always from west to east in the Copernican astronomy, though Jupiter sometimes moves from east to west according to Aristotle and Ptolemy.

Sarpi   It seems to me that no matter which way Jupiter was moving, the satellites would move from west to east in half of their orbits and from east to west in the other half, so that the choice of terms is arbitrary.

Salviati   Rather, it remains indefinite unless we specify which half of the orbit we mean. Seen from Jupiter, about which the satellites revolve, the direction is constant. Seen from the earth or the sun, about which Jupiter travels, that direction is the same as that of Jupiter's proper motion. Hence we say it is from west to east, as it is also in the upper, or more distant, half of each satellite orbit. That direction is not reversed in the nearer or lower half of the orbit, though our term for describing that motion in the lower half is the opposite of our term for the motion in the upper part.

Sarpi   Of course, and now I recall that the same point was disputed between our friend and the feigned Apelles, now known to have been the Jesuit Christopher Scheiner, in the matter of sunspots.

D&O 91 Apelles would have it that they move from east to west, the nominal direction in the lower part of their travel that we see, whereas our friend declared the motion to be from west to east in the manner that it would be described as constant by an imaginary dweller at the center of the motion.

Sagredo   I think we had better leave such questions, which relate to the system of the world, until we have finished our reading, especially as we are now approaching the end.

☆

[93]

This night for the first time I wished to observe the progress of Jupiter and of its neighboring planets in longitude along the zodiac, relating this to some fixed star; and in fact a fixed star was seen to the east, distant from the easternmost planet eleven minutes and rather to the south of it, in the following way:

(69 *bis*) ❈   ⬭*   ✳   West
East

    ✳ fixed star

    On the twenty-seventh, at one hour forty [here the printer has put "four"] minutes after sunset, the stars appeared in this arrangement:

(70) East ✳   *⬭   ✳ ✳   West

  𝒜ₓₙ ✳ fixed star

The most easterly was ten minutes from Jupiter; that closer to Jupiter, thirty seconds. The next westerly was two minutes thirty seconds away, and the westernmost one minute from that. Those closest to Jupiter appeared small, especially the easterly one, while those more distant were easily visible, particularly the westernmost, and they followed exactly a straight line along the path of the ecliptic.

    The [general] easterly progress of these planets is clearly seen by reference to the fixed star mentioned earlier, as may be seen in the diagram. But at the fifth hour the easterly star near Jupiter was one (71) minute away from it.

Sarpi   Here, of course, our friend refers to the general progress of the satellites from west to east, which we discussed a moment ago. By relating it to a nearby fixed star, any lingering doubt or confusion is removed. I now see more clearly, Sagredo, why you said this matter bears on the system of the world.

Sagredo   Yes, for whatever introduces greater regularity, uniformity of order, and simplicity of description into that system is much to be desired. Well, the text proceeds:

    On the twenty-eighth, at the first hour, only two stars were seen; one easterly, nine minutes from Jupiter, and one westerly, two minutes away. They were plainly visible and in the same straight line, perpendicular to which line, under the easternmost planet, the fixed star now fell, as in this figure:

(72) East   ✳   ◯ ✴   West

✳ fixed star

But at the fifth hour a third starlet was seen two minutes east of Jupiter, in this position

(73) East   ✴   ✴ ◯ ✴   West

✴ fixed star

On the first day of March, forty minutes after sunset, four stars were seen, all easterly, of which the nearest to Jupiter was two minutes from it; the next, one minute from that; the third, twenty [the printer put "two"] seconds farther; and from that the easternmost was distant by four minutes, and was smaller than the others. They were nearly in a straight line, only the third (counting from Jupiter) being a bit northerly. The fixed star now formed an equilateral triangle with Jupiter and the easternmost planet, as in this figure:

(74) East

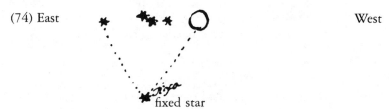

West

fixed star

On the second, thirty minutes after sunset, there were three planets, two to the east and one to the west, in this arrangement:

(75) East

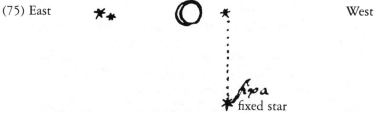

West

fixed star

The easternmost was seven minutes from Jupiter and was separated from the next by thirty seconds, while the westerly one was two minutes from Jupiter. Those at the ends were brighter and larger than that in between, which looked very small. The easternmost seemed a bit raised to the north from the line through the others and Jupiter. The previously noted fixed star was eight minutes from the westerly planet, along the perpendicular through the latter to the line passing through the planets, as shown in the figure [above].

I have reported these relations of Jupiter and its companions to the fixed star so that anyone may understand that the progress of these planets, both in longitude and in latitude, agrees exactly with the movements [of Jupiter] calculated from tables.

Such are the observations concerning the four Medicean planets recently discovered by me. Although from these data their periods have not yet been determined in numerical form, it is legitimate at least to put in evidence some facts deserving of notice. Above all, since they sometimes follow and sometimes precede Jupiter, by the same intervals, and they remain within very limited distances either to the east or to the west of Jupiter, accompanying that planet both in its retrograde and direct movements, in a constant manner, no one can doubt that they complete their revolutions around Jupiter and at the same time effect, together, a twelve-year period about the center of the universe.[27]

Salviati   How far our friend was from right in supposing that no one would doubt the rotation of the Medicean stars around Jupiter, we all know. There is a vast difference between confidence grounded in repeated observations and careful thought, and the confidence of philosophers in metaphysical principles that transcend all experience of mere astronomers.

Sagredo   The wonder is that philosophers grant even planets to exist, and do not attribute belief in them to vulgar illusions of our senses. For I do not see how the manner of reasoning about Jupiter by astronomers differs from our friend's way of reasoning here about its companions. At any rate, he goes on:

☆
[95]   That they also revolve in unequal circles is manifestly deduced from the fact that at the greatest elongation [of any of them] from Jupiter it is never possible to find two of these planets conjoined [there], whereas in the vicinity of Jupiter they are found gathered two, three, and sometimes all four together. It is also seen that the revolutions are swifter for those planets that describe smaller circles about Jupiter, since the stars closest to Jupiter are usually seen to the east when on the previous day they appeared to the west, and vice versa, whereas the planet that traces out the largest orbit appears, on careful examination of its returns, to have a semimonthly period.

Here we have a fine and pretty argument to quiet the doubts of those who, while accepting with tranquil mind the revolutions of planets about the sun in the Copernican system, are mightily disturbed to have the moon alone revolve around the earth while accompanying it in an annual revolution about the sun.[28] Some have believed that this structure of the universe should be rejected as impossible. But now we have not just one planet revolving around another while both run through a great orbit around the sun; our own eyes show us four stars that wander around Jupiter as does the moon around the earth, while all together trace out a grand revolution about the sun in the space of twelve years.

Sarpi   You remarked earlier, Salviati, that our friend referred but once in this book to motion of the earth, in a passage you had forgotten. But here he has called the earth a planet, by the word "another." So it is clear that in his book on the system of the world he intended to follow the Copernican astronomy.

Salviati   Probably you are right, though it does seem strange that his hints on so important a matter should have been left so obscure. I was not alone in overlooking them, for the published attacks against this book were mainly devoted to other matters.[29] Here he seems to me to remove an objection, not to assert a belief, and the word "another" occurs in citing the objection, against which our friend writes ad hominem.

Sagredo   Again we are running ahead of our task, which is first to finish the reading of this book. There is not much more, so I proceed:

☆   And finally we should not omit the reason for which the Medicean stars appear sometimes to be twice as large as at other times, though their orbits about Jupiter are quite restricted. We certainly cannot seek the cause in terrestrial vapors, since Jupiter and its neighboring fixed stars are not seen to change in the least while this increase and diminution are taking place. It is quite unthinkable that the cause of variation should be the changes of distance from the earth at the apogee and perigee [of the Medicean stars], for by no means could a small circular revolution produce this effect, and an oval motion—which in this case would have to be nearly straight—seems unthinkable and is quite inconsistent with appearances.[30] But I shall gladly explain what occurs to me on this matter, offering it freely to the judgment and criticism of thoughtful men.

It is known that interposition of terrestrial vapors makes the sun and moon appear large, while stars and planets are made to appear smaller. Thus, the two great luminaries are seen as larger when they are close to the horizon, where stars appear smaller and for the most part barely visible. And thus stars appear quite feeble by day, and in twilight, but the moon does not, as has just been said.

[96]

Now, from what has been said above, and even more from what we shall say at greater length in our *System*, it follows that not only the earth but also the moon is surrounded by an envelope of vapors; and we may apply precisely that same judgment to the rest of the planets. Hence it does not seem entirely impossible to assume that around Jupiter there also exists an envelope denser than the rest of the aether, about which [envelope] the Medicean planets revolve as does the moon about the elemental sphere [of earth and air.] By reason of that envelope they appear smaller at apogee and look larger when they are at perigee by removal, or at least attenuation, of the said envelope.

Time prevents my proceeding farther, though the gentle reader may expect more, soon.

THE END

Sarpi   Thus, at the conclusion of the book, we have once more the source of my principal question. In spite of this promise to write on the system of the world, and soon, he has given us little more than the periods of the satellites, published at the beginning of his book on bodies in water. Yet we know that he long ago preferred the Copernican system, having assumed that in order to explain the tides. Although he hardly mentioned that system in the book we have just read, he predicted its victory in his last book, on sunspots. Surely after such long reflection on it, his system is evident; yet he has not kept his promise of three years ago.

Salviati   When you opened our discussions with your question about this long delay, Fra Paolo, I assumed, as I believe we all did, that the promise related only to a choice among the rival systems debated among astronomers on purely astronomical grounds. Accordingly I undertook to explain the delay by telling you of work that has occupied our friend since his move to Florence, required partly by the problem he had just presented to all astronomers, partly by new discoveries, and partly by objections and oppositions to those he had already announced. Your hints about his new science of motion and his explanation of tides, however, suggest other sources of delay, even more cogent perhaps, concerning which I still lack information. I shall accordingly proceed with what I had in mind at the outset, if you are content, leaving other and more interesting speculations to the end.

Sarpi   Very well; please start from the beginning.

Salviati   The first thing that concerned our friend was to find exact places for Jupiter's companions in the world system, before publishing a book about that. The work was interrupted by his move to Florence. By the time he had arrived at approximate periods for all four satellites, attacks against his discoveries were appearing. Some of these, unpublished, were already mentioned yesterday. Others were soon printed by a Bohemian astronomer, Martin Horky, and a young Florentine, Francesco Sizzi. Accordingly our friend wished to proceed with circumspection and to improve his first approximations. In the course of doing this he found new things of interest, and was also confronted with a new attack, philosophical in character, by Julius Caesar La Galla, a professor in Rome. Perceiving the extent and variety of oppositions against his work, and at the same time recognizing that he had much more to present than he had anticipated when making this promise in his book, he deferred its fulfillment, and still does so, for reasons that I think you can discern.

Sagredo   I have the books you mentioned in my library and saw nothing in them worthy of reply. Horky's ignorant libel was more than adequately answered by a pupil of our friend's shortly after he left for Florence. Of Sizzi's book I read only the opening pages, which were too silly for words, and La Galla's book was so aridly philo-

	sophical as to have no bearing on true science, at least in my opinion. Did such trifling books really alter our friend's plans?
Salviati	The first you name did not, after the reply to it by John Wedderburn. But there was something in the second that I should call to your attention, since it came near the end, which you say you did not read. It bore directly on the observations we have just read, and it will also offer a convenient bridge to the account I must give you of our friend's next studies. As for La Galla's book, that could safely be ignored by you, who did not intend to publish on the system of the world, but not by our friend, who does, and who must overcome the kind of objections raised by professors of philosophy if he hopes to win their hearing for his system.
Sarpi	Let us next hear, then, what it is in Sizzi's book that Salviati regards as important. Since I have not read it, Sagredo, tell me why you laid it aside after reading only a few pages.
Sagredo	Sizzi began by arguing that there can be no wandering stars in excess of seven, the number accepted in past ages. He even asserted that Sacred Scripture so declares in many places.
Sarpi	Then he must be a subtler theologian than I, for I cannot recall in the Bible any mention of planets, except indirectly of Venus as the morning star. Did this subtle author object against the assumption that the earth moves?
Salviati	Sizzi mentioned that three or four times, but only in passing, and he seems to have considered that hypothesis acceptable. It was mainly against increasing the number of wandering stars that Sizzi directed his attack, assuring our friend repeatedly that he wrote not to deride his conclusions, but only in search of truth.
Sagredo	Even though it takes us a bit afield, I want to know what our friend said about that.
Salviati	He was in Rome soon after Sizzi's book came out and was told by Jesuits there that it was the most ridiculous book ever printed. Friends urged him to reply to it, but he instead apologized for its author and said that he had rather have Sizzi for a friend than anger him by replying to various inanities in the only way that such

exposures can be written. At first I was puzzled by this, but later events showed him to have judged rightly. Sizzi is now living in Paris, where he and some good mathematicians have been won over to our friend's views by his last two books, and he has sent to Italy some significant observations about sunspots carried out in France.

Sagredo   If Sizzi was won over by our friend's last books, I should not obstinately refuse to hear whatever it is that you think deserves our attention in Sizzi's book. What induced our friend to abstain from replying to it must have been something I did not read, which he took as evidence that Sizzi would in time perceive and correct his own errors. Please excuse me, then, while I go to find my copy of his *Dianoia*.

Sarpi   While our host is away I shall make a guess about what we are going to hear, Salviati. You said that this was closely linked to the satellite observations we have just finished reading, and you implied that what Sizzi challenged was their real existence as wandering stars. We ourselves noted during our reading that there were some puzzles and inconsistencies among the observations. I shall guess that Sizzi noticed and invoked those as evidence against the reality of the satellites. That would explain your further statement that the discussion will be a good bridge to our later consideration of work done by our friend since he left Padua.

Salviati   I should not have thought it possible, Fra Paolo, that anyone could deduce so accurately what we are about to hear, having only my deliberately cryptic remarks to work from. Not that I meant to conceal anything; I spoke as I did only to avoid wasting time.

Sarpi   In affairs of state I have learned to listen carefully to people's exact words, to guess as well as I can what lies behind them, and to take into account any possible motives for puzzling behavior. Puzzled by our friend's forbearance under attack, even when others urged him to expose the young man's inanities, I have sought to explain that.

Salviati   No time remains for me to ask how you deduce Sizzi's youth, because I see Sagredo now returning with the *Dianoia*.

Sagredo   Here is the book, Salviati; as you can see, it has lain practically untouched, so I have also brought a knife to use in separating the remaining pages.

Salviati   While I do that, let me say that Sizzi had been present on one occasion at Florence to witness an exhibition of the telescope by our friend at the ducal court. He had therefore seen the satellites, and what he argued in his book was that they were not wandering stars but mere optical effects that had deluded our friend. Ah, now I have found a good place to begin:

iii, 228

Here is our first argument, deduced from the nature of planets. If these planets were real, and not fictitious or imaginary, they would observe a determinate law of movement and a determinate period of return to the same placements. But it is clear that they do not observe this; in fact, in the same determinate times they do not return to the same places—for example at conjunctions, mean longitudes, and oppositions—nor do they have the same motion at the same places, as evidently appears in your 65 observations set forth from page 17 to page 28.

Sarpi   It seems to me it would have been better to say "extreme" rather than "mean" longitudes, if Sizzi meant separations from Jupiter midway between satellite conjunctions and oppositions, positions in which a satellite is in line with the center of Jupiter.

Salviati   Sizzi used the terms that astronomers long ago devised for the supposed epicyclic motions of planets, before any satellites were known to exist. Your suggestion is a good one, since midway between alignments with Jupiter, a satellite is at its maximum elongation as seen by us, which might well be called "extreme longitude" measured from Jupiter's center. Likewise, instead of "conjunction" it would be clearer to say "perigee," or position nearest to the earth, and for "opposition" we might more usefully say "apogee" to denote the position in which Jupiter lies exactly between the satellite and our eyes.

Sagredo   Despite Sizzi's inappropriate terms, I can see why our friend expected better things from him, given time. The important thing was to require determinate laws of motion, as Sizzi did. He then rashly denied that such laws exist, when he should have used our friend's observations to seek regularities. But why did he note only 65 observations, when we have just counted 75?

Salviati   Sizzi counted only the diagrammed positions, whereas we have assigned numbers to some positions for which there were only descriptions unaccompanied by diagrams. In what follows, I shall alter Sizzi's numbers to agree with those that we have noted for our own use and reference. He went on:

iii, 228   In fact you conclude that the planet which moves in the largest circle completes that [motion] in 15 days, according to your corollary on page 28 when you say: ". . . the planet that traces out the largest orbit appears, on careful examination of its returns, to have a semimonthly period."

Sagredo   Again, it was rash to say that our friend concluded 15 days, as if "semimonthly" meant any exact number. Kepler, as I recall, departed still further from our friend's estimate, writing "fourteen days" in his own *Narration of Observations*. A year later, our friend

CES 18   put it as 16 days plus 18 hours, in the book we read with Professor Simplicio a few days ago.

Salviati   As you both already see, Sizzi's error began by treating as exact what our friend offered as a first rough approximation. That is a trite device of philosophers, to discredit a whole book by exposing one statement in it as mistaken.[31] To do that, Sizzi next treated our friend's published satellite positions as so precise that not the sky itself, let alone astronomical calculation, allowed any appeal from them. Instead of making proper use of all the data supplied in the *Starry Messenger*, Sizzi applied strict logic to those selections, as any natural philosopher and no astronomer would do, writing:

iii, 229   But this does not accord with your observations. In fact, for example, from 7 January, when the observations began, to 2 March, 44 days elapsed, whence in theory it [IV] ought to be found eight times, or at least seven, at the maximum distance from Jupiter, taking account of eastern and western longitudes in its mean circle. We shall consider your most favorable observations, which for the moment I shall adduce in order that your opinion may be the better defended, so that what we deny will become evident. From observation (38), made at the seventh hour on 2 February, it appears that the maximum distance of the star from Jupiter is 14', westerly. Fifteen days before or after, we do not find the same distance just when it should be; yet on 11 February, nine days after the said observation, the star appears in its mean [i.e., extreme] longitude opposite to its former place, and distant from Jupiter only 12' when it ought to be 14' distant. Nevertheless on that night three observations were made, (50), (51), and (52) within five hours.

Salviati   It was precisely observation (38), with which the comparison is here made, that contained the error I mentioned earlier as having had an interesting history. Sizzi might have detected the error by using the same reasoning as above, but assuming that IV should have remained only 12' from Jupiter at the time of both observations on 2 February. In fact it did, IV having moved only imperceptibly in the hours between (37) and (38). Three observations were made on 11 February, making its placement on that night the safer base. Nor was IV then at its maximum elongation from Jupiter, which by several observations was clearly greater than 12 minutes.

Sarpi   Looking back at (37), I see that IV was already 12' west of Jupiter, and I now notice that in order for (38) to be correct, the separation between the two westerly starlets would have had to remain unchanged at 8'. That seems most unlikely, those two satellites being swifter than the other.

Salviati   The position of IV recorded for observation (38) was simply mistaken. Our friend, having put down the separation of 8' between the two westerly stars early in the evening, carelessly repeated that figure later. Thus it appeared in his journal, and the incongruity did not strike him when he copied it into his book. In fact the first time he noticed it was when he was reading Sizzi's criticism. He wrote a note in his copy, not at this place but a little

iii, 229   later, where Sizzi repeated his remark and added incorrectly that our friend had said that IV was at its extreme longitude from Jupiter, at observation (38).

Sagredo   It is evident how this error in reporting by our friend came to have power to mislead others in their analyses. Not many of his observations put IV as distant as 14', and since (38) was the first to do so, everyone who tried to determine the period of revolution of IV started from an impossible position of that satellite.

Salviati   Exactly. It would not take an experienced astronomer long to find out that something was wrong with the position of IV in (38) by using all the other information contained in the *Starry Messenger*.

xi, 74–76   Sizzi, however, was still baffled in March 1611, when Magini sent him a nearly correct period of motion for IV. Nevertheless Sizzi could have reached the correct period for IV through the next observation described in his *Dianoia*, which he used only to deride our friend's "semimonthly" approximation, and not to find the correct period for himself:

iii, 229   On 19 February, seventeen days later, the same star (if I may call it one) is observed separated 13' from Jupiter to the west, and thus in observation (65) there is lacking 1' of the first distance. Yet you make notations as if distances are observed accurately to one-twelfth, one-sixth, a quarter, and a third of a minute. . . . Hence you wrongly deduce a fifteen-day rotation.

Sagredo   It seems to me that you were overkind to Sizzi in saying that he could have obtained "the correct period" from this, since he would take the period to be a bit more than seventeen days, and we have seen from our friend's book published later that it is a bit less.

Salviati   Technically you are right; as long as Sizzi took (38) as a base and supposed IV to have been then 14' from Jupiter, he could not get the period correctly. In a letter to Magini he added six hours

xi, 75   twenty minutes to seventeen days, to gain that supposedly lacking minute of arc. But the separation at observation (37) was 12'; at (65) it was 13', and if Sizzi had subtracted his six hours from seventeen days, he would have had the period of IV almost exactly.

Sarpi   By what reasoning could Sizzi have subtracted the time that in fact he added? It seems to me that he reasoned correctly, and that it is mere accident that a nearly exact conclusion would have followed the other way.

Sagredo   No, Fra Paolo, I see what Salviati means. Sizzi was intent on proving that fifteen days is *not* the period of IV. That could be proved either way, since fifteen days is neither 17 days six hours nor 16 days 18 hours. The latter is, however, very nearly the period of IV, and the question is whether Sizzi might easily have discovered that. Salviati says he might have, if instead of deriding our friend's published estimate he had used all the data in search of a better estimate. For from observations (37) and (38), the position of IV on 2 February was either 12′ or 14′ from Jupiter. Seventeen days later it was 13′, confirmed by three separate findings. Hence seventeen days is either six hours too long or six hours too short. Which it is can be determined by using other observations of IV when it was very distant from Jupiter, since the alternative periods, differing by half a day, would have consequences so different as to decide the question.

Sarpi   Now, that is very interesting. You mean that a gross error in the data can be easily detected if there are enough observations, so in this case error could even be turned into a clue to the truth, if Sizzi had been interested in finding that instead of proving an inconsistency.

Salviati   That is so, and it shows the advantage of reducing scientific problems to careful measurements when possible, as astronomers have always done. Natural philosophers just go on talking about debatable qualities and assumed causes, with no real advance in knowledge of nature. As long as Sizzi followed their example he learned nothing, but before long he was won over to the true path of science, as I told you.

Sarpi   I begin to see, with Sagredo, how our friend perceived that Sizzi would eventually do so, and hence forbore to attack his book. Despite its initial reasoning from the number seven, it did contain a serious examination of some observations and measurements, of which we have heard the opening part. That section began by assuming correctly that if the things seen were satellites, they would observe regularities of motion around Jupiter. Our friend foresaw that from that assumption and full examination of all the data, Sizzi would find how to correct his own initial errors.

Salviati   Yes; and for the same reason in reverse, our friend did not waste time replying to those others, who never seriously examined the measurements but argued only from philosophical principles. Without replying to any opponents, our friend quietly went on recording measured observations when he could, sure that that would be the key to determining the individual orbits and periods of all four satellites.

Sagredo   You promised to explain how he eventually solved that problem, which he said in his next book was done at Rome about the time when Sizzi's book was published. It seems to me that you should do that now, unless there are other things in Sizzi's book that should be discussed first. The first part, which I have read, is not

iii, 212–14   worth our bother with its pretended proofs that the number of planets must be exactly seven. But Sizzi did have some things to say that bear on real problems with our friend's first observations, so he may have had other arguments that deserve our attention.

Salviati   What we have just read is followed by several further conflicts among observations recorded in the *Starry Messenger*, but I believe that it will suffice for our purposes to stop at these. Still other parts of Sizzi's book can be discussed, if necessary, together with the arguments used by Horky and by La Galla, as parts of the Peripatetic attack against our friend's discoveries and the conclusions he drew from them. We could turn to those next, or take up our friend's solution of astronomical problems presented by Jupiter's satellites.

Sarpi   I had two reasons in suggesting the latter. The first reason is that we are already familiar with philosophical oppositions to the new discoveries, and they are of interest to me only as they bear on our friend's delay in writing on the system of the world. The second reason is that if we take up the objections now, it will be more difficult later to recall the record of observations and to resume the

Sagredo   thread of our present inquiry into the problem of satellite motions. I agree with Fra Paolo. Moreover, after we have learned of our friend's progress it may be easier for us to assess the philosophical arguments raised in ignorance of his subsequent observations and reasoning—if indeed such arguments deserve our consideration at all.

Salviati   Then I shall next outline the steps by which our friend determined the periods and orbits of the new starlets, in the course of which he transformed his telescope into a measuring device of unheard-of precision. But we have already talked enough for today, and I can best present what I have to say next if I first review some notes I brought with me and make a few calculations.

Sagredo   Fra Paolo and I will leave you to this, then, and walk together to his convent, returning with Fra Fulgenzio, who will join us at supper here tonight.

Salviati   Good; I shall begin at once. An hour or two will suffice, at the end of which time I shall expect your return with our witty and intelligent companion of yesterday evening.

End of the Second Day

THE THIRD DAY

Salviati Yesterday we completed our reading of the new scientific discoveries announced in the *Starry Messenger* and considered some objections raised against them. Others remain to be discussed, but first I think it best for me to explain how our friend solved the problem he had proposed to all astronomers, which was to determine the periods of all four starlets circling around Jupiter. In case you have forgotten, Fra Paolo, those were given at the beginning of our friend's book on bodies in water that Sagredo and I recently read with Professor Simplicio. It was published last year

and stated the periods in days, hours, and minutes. Our friend wrote that he had found all four while he was at Rome in April, 1611, though of course he had subsequently improved those first determinations. Doubtless you will want to know how he solved

the problem and then went on improving the accuracy of his calculations, because the time and effort required for that had much
to do with his long delay in publishing on the system of the
world.

Sarpi   Indeed I do want to learn of this, but what you have just said
stirred in my mind a certain question having to do with the very
nature of science. I wish to put this question first, if I can manage to state it clearly. Let me try.

Sizzi found some positions of the starlets reported by our friend
that could not be reconciled with uniform circular motions
around Jupiter, from which he concluded that they could not be
truly heavenly bodies. Meanwhile Kepler, having made some observations of his own, stated publicly that although the outermost
satellite makes fortnightly revolutions, the periods of the others
iii, 184   might never be known. Now, Kepler is unrivaled among living
astronomers in resourcefulness and patient calculation. How,
then, can we say that our friend announced a *scientific* discovery
when, on the very title page of his *Starry Messenger*, he asserted
that four new stars had been found circling around Jupiter?

Sagredo   I do not quite understand your puzzle, Fra Paolo. Surely the four
periods he announced last year served to "save the appearances,"
as philosophers describe the only proper concern of astronomers.

Sarpi   Yes, it is *now* a scientific statement to say that four starlets circle
Jupiter. But was that a scientific discovery when our friend first
announced it, or just a fortunate conjecture? The term "scientific"
has always been reserved for conclusions known certainly to be
true, deduced by rigorous reasoning from unquestioned principles. In order to preserve the meaning of *science* long accepted by
all philosophers, it seems to me that Salviati should not have said
"scientific" discoveries were announced in the *Starry Messenger*,
but rather only some new astronomical hypotheses.

Salviati   I see your point, Fra Paolo, but I think the question you raise has
not so much to do with the nature of science as it has to do with
the propriety of certain ways of talking, or I might say with the
nature of language. In one way our friend's conception of science,

which I accept, is novel and disturbs philosophers; yet in another way it is old, established, and traditional. We should not wander too far from our original purpose, and I believe that by proceeding as I had planned, light would be thrown on your puzzle, as Sagredo calls it. But a short digression into the origin and nature of astronomy may be useful, so let me ask you, Fra Paolo: Do you consider astronomy a science?

Sarpi   What an odd question! Of course I do; it is the oldest and noblest of sciences.

Salviati   Then how about cosmology, as sketched by Plato, made geometrical by Eudoxus, and modified by Aristotle?[32]

Sarpi   I see what you mean; cosmology is an even older and nobler science. The science of astronomy grew out of it, to save the appearances, as Sagredo said a moment ago.

Sagredo   Excuse me, Fra Paolo; what I said was that philosophers so describe the sole duty of astronomers. My own view is that nothing stands less in need of saving than actual phenomena, which is merely Greek for "appearances." To me it appears that astronomy has been used to save the theories of philosophers, not the phenomena of nature. If cosmology is properly called science, Fra Paolo, astronomy would better be called something different, since the two did not originate in the same way, employ different methods, and have different criteria of truth.

Salviati   Our friend would approve your suggestion of using different words to describe cosmology and astronomy, Sagredo, since the two do not proceed alike. As a lover of simplicity and efficiency, however, he would avoid coining a new word and would use words we already have, calling astronomy "science" and cosmology "philosophy." He often says that science will enable men to philosophize better, which of course it cannot do if it continues to submit humbly to the jurisdiction of past philosophy.

Sarpi   Gentlemen, I was responsible for this digression and I thank you for what you have said. That is enough to show me how my difficulty arose from assuming that science must embrace all certain knowledge and nothing else. Already I see that to broaden that scope might make science more serviceable to mankind. Discoveries that enable philosophers to improve their theories, instead of merely saving them (as Sagredo put it a bit sarcastically), would then be properly called scientific, even if not reasoned from unquestionable principles. Now, Salviati, I am ready to hear how our friend solved the problem that Kepler himself deemed unsolvable.

Sagredo I am no less anxious to hear that; but before you begin, Salviati, I want to set forth a speculation of my own about how our friend proceeded, and indeed must have proceeded, in my opinion. Last night I fell to reflecting on the numerous and perplexing observations that we hurried through yesterday, and I hit upon a way of attacking the problem posed by our friend. It is so clear and straightforward that I am surprised that it seems not to have occurred to Kepler.

Salviati If what you are about to say is what I think it will be, Sagredo, I am fairly sure that it did occur to Kepler and that our friend thought so too. Finally, I can show it to be not the attack employed by our friend, for the very same reasons that probably made Kepler dismiss it from his mind.

Sagredo Well, that is a most astonishing prediction, Salviati, and it seems to me that we should make a little game of this. If I understand correctly, you think you already know what I am about to say, so please write it down now; here is paper and a pen. When I am through speaking, we will read what you expected, and if I have said the same, you shall tell us how you arrived at your further conclusions about Kepler and our friend.

Sarpi A charming suggestion, Sagredo. If only the customary acrimonious disputations among professors would give way to friendly challenges of this kind, how much more could be learned from them, and how much more entertainingly!

Salviati I have written down the plan that I believe you will propose, Sagredo; please proceed.

Sagredo From the *Starry Messenger* we learned that the outermost satellite moves as far as 13 or 14 "minutes" from Jupiter and returns about semimonthly. The outermost alone can be positively identified, because near its extreme positions it separates itself far from any of the others. Accordingly we could, by continued observations, fix the time of one such position and the period of return for the outermost starlet. Assuming its motion to be circular and uniform, that outermost satellite could be eliminated from every observation in our possession. This done, satellite III would be the outermost of the remaining satellites, separating itself from the rest and permitting us to find its maximum elongation and period in the same way. Eliminating III, satellite II would in turn become "outermost," and with determination of its period and orbit the problem is solved.

Sarpi   The reasoning is clear and sound, though (since I am no astronomer) I must ask how the outermost satellite is to be eliminated from positions near others. All four look alike, so would they not be easily confused when near Jupiter?

Sagredo   Yes, but we assume each motion to be circular and uniform. Knowing the time of one position and of one complete circuit for each accordingly suffices. We see each satellite as if it moved non-uniformly along a straight line through Jupiter. A table of sines (or chords) is all we need for calculating any seen position along that line, using the time elapsed after one maximum elongation.

Sarpi   Of course; now I understand your plan, though I also see a difficulty. Please read to us what you wrote down, Salviati, before I speak of that.

Salviati   As I expected, Sagredo reasoned quite correctly how to solve the problem "in principle," as we say. I briefly summarized the procedure thus:

> Using the maximum elongations of IV and its period of return, eliminate IV from all past diagrams of observations. The maximum elongation of III will then be seen, and its period similarly revealed. Eliminate III as done with IV, and the complete solution is evident.

Sagredo   The chagrin I might perhaps feel at having been anticipated is

	eclipsed by this evidence that my reasoning was not defective, and even more by your having obliged yourself now to answer my previous questions. But Fra Paolo has an objection to enter first.
Sarpi	Not an objection, Sagredo, but a possible difficulty. Among our previous observations there would be many in which the satellites would be grouped close to Jupiter, and since the estimated period of return could be only approximate in each case, the wrong satellite might sometimes be eliminated. That, it seems to me, would greatly perplex the further analysis at each stage. Even a single incorrect record entirely misled Sizzi, when he took a very approximate period as if it were exact.
Sagredo	Not only are you right, but I now see that some incorrect eliminations would be almost inevitable. Sometimes two satellites would be so close together as to be seen as one, and a perfectly correct cancellation would result also in a wrong one.
Sarpi	Add that two satellites must be seen as one whenever they happen to be in line, as when one is passing another. There would, then, be many uncertainties in the analysis of as large a number of observations as would be needed to give us any hope of success.
Salviati	There is still another, and more formidable, difficulty that I shall explain presently. Well, you have already seen why correct reasoning, though that alone sufficed for Aristotle in solving metaphysically all the complex problems of cosmology, is not sufficient—though it is very necessary—in solving one simple problem of astronomy. I wonder, Sagredo, if you still want me to answer your questions concerning this matter?
Sagredo	Only partly. I now see why the scheme I proposed would probably have occurred at once to expert astronomers and how they foresaw difficulties in carrying it out successfully so great as to make it hardly worth the time and effort of attempting. But I should still like to know why all that you said seemed so certain to you.
Salviati	First of all, Sagredo, your reasoning was not only correct, but so far as I know, yours is the only simple line of perfectly rational attack on the problem. Accordingly, it was reasonable to assume

that the resourceful astronomer Kepler spoke as he did not for lack of any idea how to proceed, but because he saw only one way—yours—and despaired of its success. Otherwise, if he had found hope either in it or in some alternative attack, he would have remained silent while he confidently worked to solve the problem our friend proposed. It would be no small triumph if Kepler, not having made the sensible discovery, nevertheless first transformed it into a fully scientific discovery. Second, as to our friend, I happened to know what he did the very next day after he found the clue that led him to solution of the problem. Should I tell you that now, or later when I explain that clue?

Sagredo   Now, please, for that will complete your answer to my question.

Sarpi   And for another reason, because something you have said continues to puzzle me and I want to return to it.

Salviati   Very well. Writing the next day to the ambassador at Prague about his discovery of the phases of Venus, our friend added at the end:

x, 483   "I hope to have found the method of determining the periods of the four Medicean planets, regarded with good reason by Signor Kepler as unexplainable." The words "with good reason" surely meant that our friend credited Kepler with having exhausted any purely rational schemes before he despaired in print of solving the problem.

Sagredo   I am entirely satisfied. What is your puzzle, Fra Paolo?

Sarpi   A moment ago, Salviati distinguished between cosmology and astronomy by saying that correct reasoning from first principles is all that is used in the former, but not in the latter. I want to know what else is required in solving problems of astronomy.

Sagredo   I supposed he must be alluding to more careful observations, accurate measurements, and mathematical reasoning; is that not correct, Salviati?

Salviati   Partly, though you should have said "a certain kind of mathematical reasoning," because Aristotle's cosmology was built on mathematical reasoning by Eudoxus even more subtle than Ptolemy needed for his astronomy. However, you left out the main thing I had in mind, Sagredo. What has no place in cosmology, or in any other part of philosophy, but is very often needed for solving scientific problems worthy of the name, Fra Paolo, is luck.

Sarpi   Surely you are joking, Salviati. Problems of science are not considered solved except by necessary and inescapable demonstrations; in that respect they are exactly like the problems of philosophy.

Salviati   Not exactly, because the propositions of philosophy are not obliged to describe the imperfect visible world around us. Problems of

CES 10   science, as our friend defines them, are those which involve sensate
D&O 182   experiences as well as necessary demonstrations. To solve them one must therefore usually have a bit of luck; for though we can control the demonstrations we adduce, we cannot control the sensate experiences that nature adduces, and to make the former fit the latter we must first hit upon some clue that is not evident by reason alone.

Sarpi   Salviati, I can believe that you are trying to answer my question conscientiously, but I cannot believe that our friend depends on good fortune for the advance of science. Yet I admit that he does seem to have been uncommonly fortunate in finding new truths. I must be missing something in what you wish me to understand.

Salviati   When you say he does not depend on good fortune, Fra Paolo, you mean that our friend does not just sit idly, waiting for luck to favor him. That is certainly true; no one I know thinks more incessantly or works harder. The sense in which science depends on good fortune is this. Nature exhibits a marvelous coherence and consistency in her phenomena. Exact coherence and consistency exist also among propositions of mathematics. With luck, some phenomenon of nature may be connected with some proposition of mathematics. We can in such cases say with absolute confidence that other necessarily connected propositions of mathematics will lead us to other necessarily connected phenomena of nature. But to find any connection at all between any of the infinitely many phenomena and the infinitely many mathematical propositions, or at least to find any new connections, not established long ago by Archimedes, or Ptolemy, or other great observers and mathematicians, one must have remarkably good fortune. Pure speculation, and even correct reasoning from unquestionable principles established centuries ago and universally granted, have been notably unproductive ways of going about this business. Our friend has a more fruitful way, as evidenced by his uncommonly good fortune that you mentioned a moment ago. Does this help you to understand what you seek to know from me?

Sarpi   It certainly does, and though it seems to me that you use words like "luck" and "good fortune" in rather strange ways, you have given me some interesting things to think about. I suspect that you spoke as you did because you know what is coming and that it will throw a clearer light on what you have chosen thus far to phrase somewhat cryptically. So now I beg you to tell us how our friend did succeed, solving in one year the problem which other astronomers had not solved after two years.

Sagredo   Yes, please do—after you have told us about that other, more formidable, difficulty in my scheme for finding the satellite periods.

Salviati   This is that your plan required us to start from a time of maximum elongation. At such positions it happens that a satellite—and especially IV, from which we should have to begin—appears to move along the line through Jupiter so slowly as to seem motionless for hours, or even a day. It really moves uniformly in a circle, but most of its motion at such times is directly toward us or away from us. Hence of all possible positions to be taken as our "epoch," or starting point for calculations, the most uncertain of all as to exact time of occurrence is precisely the position required to be determined first in Sagredo's purely rational approach to the problem.

Sagredo   As Kepler would have known from his countless planetary calculations, in which epicycles are exactly like the satellite motions. And of course our friend would have known the same from his already numerous satellite observations.

Salviati   Yes. But while Kepler saw no way around this and the other difficulties, our friend was confident that in time the satellites would betray their secret of coherent and consistent behavior, and he prepared to catch them in some indiscretion. I want you now to see why, for a considerable time, he was content merely to watch.

First, I ask Fra Paolo whether the number of positions in which all four satellites are seen is infinite, or limited?

Sarpi   The number of sensibly distinguishable positions of that kind must be limited, though very large. I think I see why you chose those; because when all four satellites are seen at once, there is no need to conjecture about the identity of any not seen, or where it might be. Since the number of distinguishable positions of this kind is finite, one may expect that with time and patience the same position would be observed twice, with a long period of time between them. And long before that, two positions might be observed that were alike in so many respects that some plausible hypothesis could be formed about the periods of all four, or at least more than one, of the satellites. Dear me, you are compelling me to admit that an element of luck was present in our friend's tactics (if they were in fact of this kind)—but a peculiar kind of luck, in which something very improbable is, as it were, bound to happen. Oh! now I see that this is bound to happen only because of that universal assumption, stated by Sizzi in his objection, that celestial motions are perfectly circular and uniform. So this strange kind of inevitable luck is, after all, rooted in an unquestionable principle and would qualify as science. Excuse me for ruminating; please go on.

Salviati   The trouble with Sagredo's purely rational scheme was that one might expend an enormous amount of time and effort without solving the problem. The advantage of the program you have just outlined, Fra Paolo, is that very little time or effort need be spent until the lucky chance occurs. Even if it never does, there is no great loss, while if it ever happens, the gain will be enormous in reducing the amount of calculation, trial, and error necessary in all such inquiries. Well, now, to carry out such a program, Sagredo, what procedure and method would you recommend?

Sagredo   The very opposite of what I formerly considered the only method available. Instead of imagining that I could fix any position of one satellite accurately, let alone find its period of revolution, I would maintain careful records until two positions of all the satellites, of a kind just designated by Fra Paolo, were found—either exactly alike, or similar in some way that suggested a hypothesis of motions based on the phenomena observed. Until I found some such happy pair among those I had already recorded, I should waste no time in theorizing and calculating. My only effort would go into recording observed positions as accurately as possible until such a pair was found. That would be the first time it would pay to get down to hard work.

Salviati   And what would you record while waiting?

Sagredo   The same things our friend noted in his book; that is, times, measurements of separation, and any differences in brightness—for it might be that one of the starlets is always brighter than the rest, and another is always fainter. To be able to identify even one particular satellite from night to night would be a great advantage.

Sarpi   From the book we learned that any satellite appears faint when seen close to Jupiter, so there would be little hope of gaining such an advantage. Still, the reward would be considerable, so if there were a simple and efficient way to record apparent brightness, it would be worth doing.

Sagredo   There is a way, familiar to astronomers, of using numbers for this, classifying fixed stars as to magnitude from first to sixth. I remember that in his book, our friend suggested using "seventh magnitude" for stars not seen usually, but bright when seen through the telescope.

Sarpi   Why not compare a satellite seen through the instrument with a star familiar and often seen, calling it "first magnitude" or "second," and so on according as those stars are called?

Salviati   You two have been working out precisely the plan adopted by our friend. To make comparisons easier in his search for identical or similar positions, he began at once to omit wordy descriptions and just record observed positions as dots along a line, representing Jupiter as a small circle. Between these he placed, below, a number representing separation in "minutes" or fractions thereof, and above each dot he put a number, usually 1 or 2, indicating apparent magnitude. I have seen his diagrams for the balance of 1610, and they were kept in this way.[33] There are not very many of them because of his varied activities, his move to Florence, and a period during which Jupiter was too near the sun to be observed. He found the very clue he needed, in the last month of that year.

Sagredo   I should be interested to learn what form it took. Was there finally a repetition of some exact position of all four?

Salviati   Not quite, since as it happened the two positions that first gave our friend something solid to work on occurred only a week apart, or rather, one hour less than a week.

Sarpi   That was indeed a stroke of luck, as the two diagrams must have appeared near one another on the same page, and so did not require a painstaking search of all past positions.

Salviati   In that respect it was a stroke of good fortune, though in another way the short interval was unfortunate. If there had been a greater lapse of time between the two, the periods of return could have been more precisely estimated. In any case our friend recognized the clue he had been waiting for. He marked the two positions with a cross, and on the very next day he commented on Kepler's pessimistic opinion.

Sagredo   But surely he did not solve the whole problem at one fell swoop and from a single clue!

Salviati   Far from it; the whole solution was extremely difficult and required what our friend calls an "atlantic labor" extending over many months before predictions of real value could be made. What did happen at once was that our friend knew that any labors he undertook would be rewarded. The two observations came on the nights of 3 December and 10 December. On the eleventh he wrote to the Tuscan ambassador at Prague, as I told you, saying that he hoped he had found how to determine all the satellite periods, which Kepler had reasonably believed an insoluble problem.

Sarpi   I am curious to know how nearly the two positions resembled one another; do you remember what they looked like?

Salviati   Yes; I saw them in his notes for December 1610, marked with a cross on the night of the tenth when he noted the resemblance. Omitting all others, they were as follows:

```
                      2           2      1      3
Day 3 hour 5          *     O     *      *      *   +
                         6     4     6      4
```

```
               2     2           2      1
Day 10 hour 4  *     *     O     *      *          +
                  8     6     4      6
```

The numbers below represented separations, counted in "minutes" as estimated from the apparent diameter of Jupiter. If we add these together, we have the east-west positions of the four satellites as follows:

3 December, hour 5:   E6 Jupiter W4 W10 W14
10 December, hour 4:   E14   E6 Jupiter W4 W10

Sagredo   It is easy to see that while the outermost satellite crossed its entire orbit, each of the others may have circled Jupiter some exact number of times and arrived back where it started. And the magnitudes placed above each starlet established III as the bright satellite, since elongation of "ten minutes" must belong to the second largest orbit. I wonder whether the other two starlets were also identified by separation, with II to the east and the innermost to the west.

Salviati   No, calculation using our friend's present tables shows that the innermost lay to the east, at its maximum elongation, while II was to the west at about half its maximum distance. But, as you surmised, each had circled back to its original place, very nearly, except the outermost, which had nearly crossed its whole orbit. Knowing that III had returned, our friend assumed that it had circled but once, while II had circled twice, and the innermost probably four times.

Sarpi   Why not just three times?

Sagredo   That would certainly be possible, but I see a reason for trying four times first. Since III circled once, half a week is the slowest return we can assign to II, and it is then tempting to assume each period to be about half of the next one outward. If that does not work out, little time and effort will have been wasted; if it does, the back of the entire problem is broken. But I do see a difficulty arising. One quarter of a week is 42 hours, and if the innermost one reached this position 42 hours later, that would be in daytime and it could not be seen. Since we start 4 hours after sunset, 42 hours more ends at 22 hours after sunset two days later; that is, two hours before sunset, in daylight.

Salviati   That is one difficulty, and there are others. Even if 42 hours is nearly the correct period, a small difference between that and the true period creates a large difference in position after several circlings. Besides, not only daylight hours, but rain or clouds, frequent in December, could make observation impossible; and although separation from Jupiter is reasonably easy to estimate, such judgments are anything but secure. Fra Paolo, can you suggest a more certain procedure, starting from this remarkable first clue?

Sarpi   I have been thinking of what you said earlier about very slow movement near maximum elongation, making the time at which that position is reached by any satellite hard to determine precisely. A consequence is that the very best position for determinations would be when a satellite comes in line with the center of Jupiter. Of course, then it cannot be seen; still, such a placement ought to be chosen as the position from which calculations will be made once the period of a satellite is known. If on some night we saw this fastest starlet near Jupiter on one side, and then hours later we saw it on the other side, we could estimate pretty closely when it had crossed our line of sight through Jupiter. But to wait for such a situation might delay things quite a long time.

Sagredo   Probably not long, if the hypothesis is right. In order to circle Jupiter every 42 hours, the innermost satellite must cross our line of sight twice in that time, or more than once a day. Each crossing would be three hours earlier than the previous one, so in one week we should encounter an occasion of the kind you want. Now tell us what happened, Salviati.

Salviati There were interruptions due to cloudy nights, but on the night of 29/30 December our friend was able to make three observations, well spaced apart, while he was also observing an eclipse of the moon on that night.[34] About midnight only three satellites appeared to view, though there had been four starlets earlier and again three hours later there were four. From these observations it was known that the innermost satellite crossed Jupiter's disk from east to west. It had therefore passed through perigee, between Jupiter and the earth, according to what we said earlier about the true motion of satellites from west to east. Our friend calculated that the center of Jupiter had been crossed about an hour and a half before midnight.

Sarpi How could he have known which starlet it was that crossed the face of Jupiter?

Salviati From the three observations it was possible to know the speeds with which the satellites appeared to move, which speeds differ greatly when the satellites are near Jupiter. The most rapidly moving one had crossed Jupiter, and that was the innermost satellite, for it is a well-known rule that celestial motions are more rapid in smaller than in larger circles. A week later, on 6 January 1611, two of the satellites crossed Jupiter, of which one was again the innermost. Looking back through his records, our friend found a similar event 554 hours before, on 14 December 1610. From this it appeared that the period of the innermost satellite might be only 41 hours, and not 42, so he prepared a rough table of future expected disappearances of the innermost satellite in January.

Sagredo Were not most, or at least many, of those to come during daylight hours?

Salviati Yes, but that did not matter so long as some could be observed. On 20 January one crossing occurred, and from that and the previous observations he determined that 42 hours was close to the true period. Using that information, coupled with his knowledge of the motion of IV and with the rule of speeds already mentioned, he identified a crossing of Jupiter by III four days later, this one at apogee. On 13 February, II reached perigee, and on 7 March there was a similar, and very long, passage of IV across the disk of Jupiter from east to west. Thus in two months our friend was in possession of useful epochs for all four satellites, nearly correct periods for two of them, and good approximate periods for the other two.

Sarpi   By "useful epochs" I suppose you mean best times from which to commence calculations, starting from apogee or perigee.

Salviati   Yes, for as you already noted, those positions can be found from
iii, 441   observations more exactly than any others. By a fortunate chance on 15 March, our friend was next able to establish epochs at apogee or perigee for all four starlets on a single night. Half an hour after sunset he saw three, one to the west and two easterly of Jupiter, all very near to it. Three hours later only the two to the east remained, and an hour after that even those had disappeared. During the following three and one-half hours none reappeared, by which time Jupiter was setting. This "great conjunction" of the satellites, as our friend calls it, gave him a useful epoch for each one on the same night, which he adopted in preparing his first tables of satellite motions.

Sagredo   Calculating future positions from a single night might reduce the labor, but would it not be better to choose either apogee or perigee as the starting point for each calculation, whenever it occurred?

Salviati   It would, and later on our friend chose apogee as the beginning of revolution around Jupiter for each satellite, at the time most accurately known. Shortly after the "great conjunction" our friend
iii, 946   left Florence on a long-planned visit to Rome to exhibit his discoveries. There, at the beginning of April, he prepared his first tables of motions of each satellite, leading to new discoveries that I shall next explain.[35]

Sagredo   I really should like to know more about this pioneer work on which they hinged.

Salviati   If I were to tell you at length, it would keep us too long, and so I shall mention only an error that beset the first tables. This chiefly affected satellite III, and because the error arose from sound reasoning, it aptly illustrates the network of observation and mathematical calculation that lies at the very basis of our friend's approach to problems of astronomy.

Sarpi   Without breaking the thread of your story, Salviati, I wish to add that from what I know about our friend's work in the science of motion of heavy bodies, his approach to problems of physics may be very similar, and perhaps his delay in writing on the system of the world is partly a result of further inquiries into possible connections.

Salviati   Your comment is most welcome, Fra Paolo, for it shows that the world does not lack ears to hear and minds to grasp that there is a great deal more to understanding the system of the world than merely choosing among the three or four rival astronomies already existing. Our friend often wonders whether it is worth his trouble to go on publishing when he sees his writings denounced not only by philosophers but even by astronomers. I have begun to glimpse the very same possibility you mention, but let us defer its discussion until I have heard more from you and Sagredo about the other work done here before astronomy attracted our friend's keen interest.

Sagredo   An interruption having been allowed, perhaps I may interject a question about what you said a moment ago. I see how science may be advanced by luck, but how can error arise in science from good reasoning?

Salviati   It is easy to see that even the best reasoning accompanied by some oversight may result in error; I have a good example of that to offer you later. The present case in part resembles that situation, though I do not like to say that here our friend "overlooked" something whose existence, or even its possibility, was not yet known to him. What he did not take into account was the fact that there can be eclipses of the satellites. In fact, at the first observation on 15 March, calculation now shows that it was III he did not see, it having already passed behind the western edge of Jupiter. He reasoned, however, that it must have been III that appeared to the west, and that what was not seen was IV. It seemed that III could not have been already behind Jupiter, because it did not reappear on the east even many hours later. Calculation now shows that III had in fact emerged a good way, but happened to be in eclipse. Not yet knowing of possible eclipses, our friend believed it must have been IV that was concealed by Jupiter, that satellite moving so slowly even at apogee that it could have remained hidden many hours, especially because IV is always faint and difficult to discern near Jupiter. As a result of this carefully reasoned error, so to speak, our friend assigned to III an epoch considerably later than correct and was obliged then to give it a period half an hour shorter than its true one. The two compensating errors continued to elude him for several months.

Sagredo   I see what you mean by an error created by good reasoning, but if you will not call its source in this case an oversight, it seems to me we should call it plain ignorance.

Salviati   We usually say a person is ignorant of something only when that thing is known to other people, who could have informed him. In this case it would be more appropriate to say that our friend adopted a plausible hypothesis in preference to one which at the time would have seemed to him bizarre; that is, that the shadow cast by Jupiter was capable of hiding a satellite for some hours. Nevertheless I accept your term "ignorant," for it throws light on the state of science in all past ages, and perhaps on a necessary condition of science at every age. What passed for science from Aristotle's time to the present has been the combination of excellent reasoning with profound ignorance, in Sagredo's sense of that word. If our friend's conception is correct, science will always remain approximate—that is, with some element of ignorance— though its uncertainties will forever diminish.

Sarpi   As Nicholas of Cusa pointed out long ago—unfortunately with little influence on other philosophers—the best we can expect to attain is *learned* ignorance. But, Sagredo, why did you say "the first error"?

Sagredo   In order that a second error, of assigning too short a period to III, should not go unnoticed. Good reasoning in science should not involve any contradiction of observations, which astronomers have always taken as the last court of appeal in their science.

Salviati   So does our friend take them, in his. But in the interplay of observation and reasoning that constitutes his science, it is necessary always to abide by the *preponderance* of evidence at any given time, good reasoning being as much a kind of evidence as are observations themselves. In this instance his *failure* to see any satellite at midnight was evidence; one might even say that it was observational evidence, however odd it sounds to call it an "observation" when something is *not* seen. When reason calls for observation of something that is not in fact observed, an explanation is sought that is consistent with whatever else is known (or assumed) at that time. A constructive example will be found later in the discovery of satellite eclipses. In the present instance our friend had to do *some* violence either to reasoning or to his past records. He chose to set aside recorded evidence rather than abandon a result of reasoning that could not be altered without contradicting the assumption of regularity in satellite motions.

Sarpi   Here my experience in matters of state stands me in good stead, for nearly all decisions in them are necessarily based on preponderance of evidence. Outside of our holy Catholic faith, it may be that probable evidence is the only guide men ever have in the search for truth.

Salviati   Our friend holds precisely that view. It takes judgment to decide preponderance of evidence when the facts are of many kinds; people who lack good judgment do well to insist upon strict deductive inference from metaphysical principles. As to the second error, Sagredo, of shortening the period of III to fit with the erroneous epoch, that was certainly (like the first error as I originally described it) the result of good reasoning.

Sagredo   So long as our friend did not alter his previous records, but simply chose among them in reasoning as he did, no harm was done. But how did he eventually detect and correct the two compensating errors? I suppose that must have awaited his discovery of satellite eclipses.

Salviati Yours is a perfectly natural assumption, but it is mistaken. The errors introduced for satellite III were detected and corrected not long after our friend's visit to Rome, and by a most interesting process that again illustrates the network of calculation and observation in his science. Having prepared his first tables of satellite motions, our friend put them to use by calculating where the starlets should have been at times for which he already had records of observations. As expected, he found need of corrections and refinements such as follow after any first approximation. In that process, he noted that as past observations receded from 15 March, disagreements between calculated and observed positions became always greater, and especially for satellite III. It was evident to him that if a satellite *period* erred by excess or defect, the resulting disagreement increased as the total time interval became greater. Yet no adjustment of the period of satellite III sufficed to eliminate disagreement, because the mistaken *epoch* still remained to create a disagreement that did not change with increasing time.

iii, 858

Sagredo I recall that our friend used to say that it was the great virtue of mathematics not only to call one's attention to the existence of error, but also to point to the source of error by revealing its kind. Here we have a good example of what he meant.

Salviati The same thing was repeatedly confirmed during the tedious work of calculating and correcting his tables of satellite motion, which was occasionally rewarded by similar insights. Ptolemy remarked in the *Almagest* that in forming their hypotheses, astronomers must attend not only to the phenomena but also to the demands of calculation. Our friend's work on his tables went slowly until a further source of disagreement was thus detected. This one indeed resulted from oversight on our friend's part, as I shall explain— unless you wish to comment on what has been said up to this point.

Sarpi I wonder why our friend continued to spend all the time and effort that must have been consumed in such calculations. It seems to me that his work done at Rome in April 1611 sufficed to show that the Medicean stars circled Jupiter in some regular way, after which the detailed calculation of orbits and periods could be left to others. Surely numerical details were not essential in order for our friend to write his book on the world system, and he did not have to delay that while seeking to perfect figures that may always remain somewhat approximate.

Salviati   It is true that the work of which I have spoken could have been left to others, and without loss of that glory deserved by the person who had found how to go about finding what Kepler had said might never be known. But we have already seen that there is much more to the system of the world than a choice among the three or four rival astronomies already existing. I shall now add that had it not been for the time-consuming calculations here in question, our friend would have found that choice less certain; and not only that, but some beauties of the world system would have remained concealed from him. On top of these beauties are utilities to navigation and cartography discovered through such calculations, our friend holding that science should be not only beautiful, but useful.

Sarpi   But our friend's choice remained the Copernican astronomy, as seen from the conclusion of his *Sunspot Letters*.

Salviati   Those letters were written in 1612; we are still talking about 1611, when our friend knew less, chiefly because he had not made, corrected, and refined the calculations in question.

Sarpi   You cannot be implying that he had not already chosen the Copernican system when he saw the phases of Venus, and probably even earlier.

Salviati   Of course I am not. But there is a great difference between merely preferring that system on one ground or another, or even on several grounds, and sincerely regarding it as certain of ultimate victory. Before publishing a book on the system of the world, one should be able not only to set forth one's grounds, but also to answer all known objections that can in any way be tested by experience. The phases of Venus were conclusive against Ptolemy and Aristotle, but

	not against Tycho Brahe. How the present calculations tended to eliminate Tycho's system will be seen in due course, if I may proceed.
Sarpi	Certainly you may, though I do not yet see how calculation as such can lead to anything really new.
Salviati	The very next correction of which I spoke may throw some light on that, and discovery of satellite eclipses will throw more light, as well as making more clear our friend's conception of science.

    I said that there had been an oversight in the preparation of the first tables, meaning a neglect of something called *prosthaphairesis* and known to all astronomers. There is no need to discuss this except to say that it names certain systematic additions or subtractions to the angular revolution of a planet. Chief among these is correction for parallax if the earth moves, changing the observer's position.

Sarpi	I take it that you mean some corrections to save the appearances after that revolution has been first calculated on the assumption of uniform circular motion around a fixed center—the earth for Ptolemy or the sun for Copernicus.
Salviati	Yes, though for Ptolemy the center of assumed uniform motion is not the earth, but an empty point called the *equant*, and for Copernicus it is not the sun but an empty point relatively near to the sun.
Sagredo	Granting that *parallax* must be considered in calculating true positions of a planet at different places along the zodiac, I do not see how that could affect satellites that faithfully accompany Jupiter at all times, for their positions are measured only relatively to Jupiter and not to the fixed stars.
Salviati	You forget that their actual motions are circular around Jupiter, while the motions we observe are seen as if they were made along a straight line passing through Jupiter at right angles to our line of sight. That line would faithfully reflect the actual motions if we on earth were at the center of Jupiter's revolution, but not if the sun is at that center (or very near to it). Always at right angles to our line of sight through Jupiter, the line along which we see the satellites

is not necessarily at right angles to the line through Jupiter and its center of revolution. Consequently we see the satellite motions along a line that oscillates back and forth as the relative positions of Jupiter, the earth, and the sun change throughout each year. Parallax is taken into account to correct for those oscillations of the line along which we measure relative distances between satellites from night to night.

Sagredo   Now I see, or think I do. But since Jupiter, the earth, and the sun actually do change their relative positions continually, no matter what astronomical hypothesis we adopt, it follows that the correction our friend overlooked at first would be necessary as well in the Ptolemaic as in the Copernican astronomy.

Salviati   So it would, and so much the more reprehensible was his oversight, though we must remember that in dealing with satellite motions our friend was venturing into territory previously unexplored by any past astronomer. You yourself, Sagredo, did not see at first why the corrections necessary to be applied in calculating the places of a planet need also be considered in calculating relative places of its satellites. This is one of those many things easy to understand once they are seen, but not at all easy to see in the first place.

Sagredo   Perhaps you will be good enough to tell us how it came about that our friend did see it, after some months.

Salviati   The oversight came to his attention after he discovered by calcu-
iii, 491–517   lation that the farther he went back from March 15, the date of his first set of periods and epochs, the greater became the discrepancy from positions he had observed and recorded. This was conspicuous in the case of satellite III for reasons already indicated. When its motions had been corrected for large errors in both epoch and period introduced in the first tables, and when smaller necessary adjustments had been made for the other satellites, he saw that there remained a systematic departure of observations from calculations that affected all the satellites equally, or perhaps I should say proportionally to the sizes of the orbits. In the autumn of 1611 he recognized that what was still needed was the Ptolemaic correction

iii, 521
called *prosthaphairesis*. Remarkable subsequent implications of that correction will be explained later. Because Ptolemy did not believe the earth moved, he did not recognize the correction as one neces-
iii, 522
sitated by parallax, or change of the observer's position. Our friend used the Ptolemaic tables, not those of Copernicus, which are somewhat different for a reason unnecessary to mention now.[36]

Sarpi
I am truly astonished, because our friend had already long favored the new astronomy as a result of his reasoning about the tides. It is true that I do not recall his again having mentioned his preference, or having discussed astronomy, from that time until he began his telescopic discoveries.

Sagredo
GAP 3–53
Excuse my interrupting, Fra Paolo, but you seem to be forgetting the great controversy he had with Professor Simplicio over the new star of 1604. That was certainly an astronomical matter if there ever was one.

Sarpi
I did not forget that, Sagredo, but my mind was on this curious matter of his preference for the Copernican astronomy, on which that new star had no possible bearing that I can see. I doubt that anyone in Venetian territory has forgotten the dispute you men-tion, but it concerned Aristotelian cosmology rather than planetary
GAP 5–16
astronomy. Never were two clever and distinguished professors more heatedly, or more amusingly, embattled than in that dispute which involved the very principles of natural philosophy. Let me accept your correction by amending my statement to say our friend did not discuss *planetary* astronomy, and finish what I was about to say. After our friend had observed the phases of Venus late in 1610, he began speaking out as a Copernican, so I should have thought that as late as the autumn of 1611 he would no longer use any Ptolemaic tables, as Salviati has just told us he did.

Sagredo
Because Salviati said that there is not much difference between the Ptolemaic and the Copernican tables of correction, nothing may be implied by our friend's choice. Perhaps he merely had the older tables conveniently at hand, and not the Copernican.

Salviati
I mentioned the fact only as of possible interest, not to draw from it any conclusion about our friend's preference between astron-

omies. In replying to Fra Paolo, I should like to stress the vast difference between preference and conviction. Certainly discovery of the phases of Venus deeply impressed our friend, and it was in that connection that he first began making his preference known publicly. Thus he wrote promptly to the Tuscan ambassador at Prague that "Kepler and other Copernicans have philosophized correctly." He did not write "and we Copernicans," but implied opinion rather than commitment, as if there were still many problems to be resolved. Presently we shall be speaking of a further discovery that may account for his much stronger statement at the end of the *Sunspot Letters.* Even that still did not commit him,

D&O 144   though he prophesied the ultimate victory of Copernican astronomy in view of the growing number of evidences in its favor.

Sarpi   We theologians well understand the need of great caution and prudence in matters of complete conviction. My experience in affairs of state taught me the same circumspection there, for doctors of jurisprudence and legal counsel always advised the withholding of final commitment whenever possible. Perhaps only in philosophy is absolute certainty found among intelligent laymen. It seems that our friend regards his kind of science as more like legal deliberation than like philosophy.

Sagredo   You could not have hit on a better metaphor, as became clear in our conversations with Simplicio a few days ago. Philosophical confidence in knowing causes is very different from scientific conviction formed when natural laws are discovered by measurement, knowing that it may be modified by future observations and measurements.

Salviati   Enough of this for now; let us return to our main subject. Correction of our friend's first tables and calculations by taking parallactic effects into consideration enabled him to obtain rather good agreement between calculations and past observations of the Medicean stars. Recalling Fra Paolo's remark a while ago, we could call that a reasonable time to have left further refinements for others experienced in astronomical calculation; but instead our friend redoubled his own efforts. Having done all he could to bring his tables into agreement with past observations, he perceived that they could be further improved only by making more accurate observations, so he fell to thinking how he might increase the precision of his determinations of distances between Jupiter and the satellites.

Up to this time, in January 1612, it had been our friend's custom to record the distance from the edge or "limb" of Jupiter to the nearest satellite seen at either side, and then the distance from such a satellite to the next outward, and so on. The distances, though stated as "minutes of arc," were still estimated in terms of the apparent diameter of Jupiter's disk on the night of observation. This system was tolerably accurate, especially after many months of practice, so he did not give thought to still better measurement until that became the only way in which he could hope to improve his tables of the motions substantially.

Sagredo   What about the method mentioned early in the *Starry Messenger*, of reducing the field of telescopic view by perforated disks placed over the outer lens?

Salviati   That was a delusion born of pure reasoning and was soon aban-
iii, 187   doned as the result of experience. The field of view is not much changed until the opening becomes so small that too much light is lost for effective observation. In reporting observations of his own, Kepler noted that opening of the "window" of his telescope increased its field of view very little, from "hardly half the diameter of the moon" to " very little more than half the moon's diameter."

Sarpi   How could Kepler enlarge this "window" at all?

Sagredo   He was using a telescope made by our friend, who had found from the first that an annoying haziness of image, and rainbowlike
GAW   discolorations around bright lights seen through lenses not per-
147–48   fectly ground, were remarkably reduced by placing paper rings
x, 278   over the outer lens, so that only the central parts were actually used. No doubt Kepler removed such a ring, but then found that the increased field was not more than half a diameter of Jupiter larger than before.

Salviati	The joke of the whole matter is that in the *Starry Messenger* the key to a correct method of measurement had already been set forth, without our friend's having recognized it as such.
Sarpi	Yes, I now remember that you mentioned during our first day's reading that he later made an instrument using the same principles as the test of telescopic power set forth in his book. His test was to draw two circles of different diameters on a wall, close together, and look at the smaller one through the telescope while viewing the other with the naked eye. When both appear the same size, the power of the telescope is given by the ratio of the circles, either in area or in diameter according as the "power" is taken to relate to magnification of surfaces or of lines.
Salviati	That became his inspiration for the new instrument.[37] On a circular disk of gray cardboard our friend drew a network of lines, horizontal and vertical, two *punti* apart, or about one-fifteenth of your Venetian inch. Through the central intersection of lines he drove a pin into a stick, so that the grid could be turned to different positions. At the other end of the stick he provided a wide ring that slipped snugly over the tube of his telescope. With the grid in place and faintly illuminated by a lantern, he found Jupiter with the telescope, seen with the other eye as if it were centered at the pin. By moving the ring to and fro, adjusting the distance from eye to grid, he could fit the disk of Jupiter exactly between the two vertical lines adjoining the pin. Finally, the grid was turned until the central horizontal line passed through the Medicean stars. Each ruling then represented exactly one radius of Jupiter, and the positions of satellites could be recorded to about one-half a radius, or ten seconds of arc.
Sagredo	This sounds almost too good to be true, so tonight we shall test it by using one of my telescopes and holding such a grid beside the end of it. Provided that the Jovian system can then be seen as if drawn on paper, I shall be satisfied that such an instrument could be accurately constructed and confidently used.
Sarpi	I am astonished and delighted by the simplicity of the device, once it has been explained. But you have only told us how to find the

Salviati
iii, 859

Sagredo

Salviati
iii, 486–87

number of Jovian radii in the elongation of any satellite, counting from the pin. How does our friend go on to measure the diameter of Jupiter in seconds of arc?

In the same way, using the measure of the distance from eye to grid and a table of sines of small angles.[38] The ratio of two *punti* to that distance is the sine of the visual angle of Jupiter's diameter as seen magnified. Dividing that angle by the power of the telescope gives the visual angle of Jupiter's disk subtended at the eye on any night. The visual size changes by as much as one-fourth over the whole approach and retreat of Jupiter with respect to the earth, usually very gradually, but sometimes quite noticeably over a week or two. But that makes no difference when the positions are always recorded in visual Jovian radii, since the elongations are magnified in the same ratio as is Jupiter's diameter. Our friend no longer records positions in minutes and seconds of arc, as he tried to do at the beginning.

I think he must use them in his calculations, however, because those must depend on angular rotation of the satellite around Jupiter.

So they do, but he has also devised another instrument, which he calls *jovilabe* (on the analogy of astrolabe), to save himself the tedious bother of trigonometric calculations whenever he compares a calculation with an observation. This instrument is graduated in degrees around the large circle representing the orbit of IV, to which he assigned a radius of 24 Jovian radii. Proportionally smaller concentric circles represent the other orbits. There is a thread through the very center, which represents the center of Jupiter. Stretching this thread to the number of degrees calculated to be the rotation of any satellite from apogee, at a given date and time of night, he notes the intersection of the thread with the circle for that satellite. Say, for example, satellite III has moved 310° from apogee. Vertically below that intersection, along the central horizontal line, are marked the elongations in radii of Jupiter corresponding to rotations around it,[39] in this example 11½ Jovian radii.

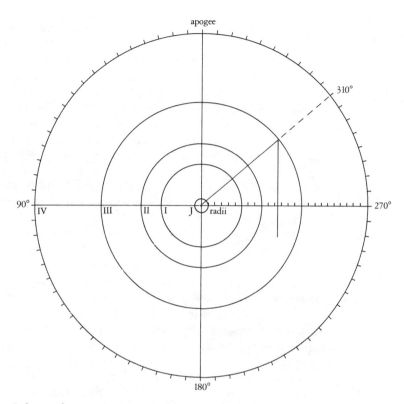

Sarpi   I do not know whether to admire this way of avoiding the tedium of angular calculations by the use of a mechanical invention or to deplore this evasion of the proper work expected of a mathematician. One thing seems certain; if our friend describes such methods in his book on the system of the world he will be laughed to scorn by philosophers, both Peripatetic and Platonist. For the former, it is bad enough to introduce mathematics into science in the first place; I remember how our famed Peripatetic Professor Zabarella[40] used to inveigh against that at Padua, as his successor Simplicio still does. They would redouble their scorn when pure mathematics is replaced by mechanical toys, and in this they would be joined by lofty-minded Platonists, who remove mathematics even farther from base artifacts of our sensible world. If our friend's actual operations ever become known, his claim to the word "science" will be rejected by every philosopher in the world.

Sagredo   It is rejected already, even without his making known his actual methods of work. A few years ago a Florentine philosopher,

GAP 80,n.   explaining the new star of 1604, called for destruction of all the instruments of astronomy, armillary spheres, astrolabes, and the rest. The telescope itself is now condemned by philosophers, and so will be all other instruments that threaten conclusions reached by syllogisms based on metaphysical principles, which cannot abide actual measurement. Well, Salviati, what resulted from these inventions of our friend?

Salviati   The first thing was a much more accurate measurement of the size of Jupiter than had ever been made before. Tycho, in agreement

D 362   with earlier Arab astronomers, placed its diameter at well over two minutes of arc. Our friend, as we have seen, treated this as one minute in his book, and finally he measured it as about 40″ in January 1612. By June it had shrunk noticeably, Jupiter being then more distant from the earth. The very considerable changes in apparent size of some planets viewed through telescopes, especially conspicuous in Venus, are easily explained by the Copernican hypothesis without huge fictitious epicycles. But let us get back to the immediate results of our friend's conversion of the telescope into a measuring device.

From February 1612 to the present, our friend's records of positions of the Medicean stars have been kept in terms of the apparent radius of Jupiter, not its diameter as before, and measured as elongations from its center, not as separations from the limb and between satellites. That alone enabled our friend to improve his tables of motion, for you recall that the exact edge of Jupiter is not seen, though the planet appears sharply bounded, and there is moreover an area of difficult visibility near it. With more accurate recording of his observations, our friend determined new epochs for all the satellites in February, and considerably refined his tables.

But at this time he was finishing his book on bodies in water for the printer, and in March a letter from Mark Welser made him realize that he must attend seriously to sunspots, so his work on the Medicean stars became mainly that of simply entering new and carefully measured observations.

Sagredo

CES 15, 196

We have already discussed the book on bodies in water, and since it would not be fair to Simplicio to speak now of the one on sunspots, let us go on to the next stage of analysis of satellite motions.

Salviati

Very well. In mid-1612, using his refined tables and including the correction discussed earlier, our friend once more undertook to make calculations and compare them with records of past observa-

iii, 517

tions. This time he began with those of mid-March 1612, his first calculation agreeing very closely with his record of that observation. But for the next night, 18 March, calculation showed that a satellite should have been seen which was not recorded. In fact, our friend had been surprised on that night not to have seen it and

iii, 449

noted in his journal that it had still not appeared around midnight, when Jupiter set and could no longer be observed. Calculation now confirmed that the satellite had been well beyond the region of difficult visibility near Jupiter's limb, and yet our friend had looked for it in vain.

The solution of the puzzle was easy; our friend wrote below the calculation: "IV was obviously in the shadow of Jupiter, for at the sixth hour it had not yet appeared." That the shadow of Jupiter ever actually eclipsed a satellite had not previously been known. Indeed, it seems rather improbable that a tiny body so remote from the sun casts a real shadow capable of having effects we can observe. Only precise calculation and confidence in his own recorded observations compelled our friend to recognize the full implications. In that he did employ method in Aristotle's sense, Fra Paolo, so you see that our friend is no mere empiricist. His initial discovery of approximate periods for all four Medicean stars had been the most perplexing problem. Construction and refinement of tables of their motions, though a herculean task, was not difficult in the same way. Our friend knew exactly what to do at each stage, and did it, just as ancient astronomers had constructed tables of planetary motions. The last and most important discovery (of eclipses) was not at all difficult when the tables of satellite motions were refined, though it could not have been made without long previous labors of two different kinds.

Now, what this discovery of satellite eclipses revealed was a necessary consequence of the Copernican motions. You need not remind me, Fra Paolo, that it was no less a necessary consequence of the Tychonian motions, because mathematically those two systems are equivalent. But what happened next shows that astronomy is no less physical than mathematical. Our friend explained to himself how a satellite comes to be eclipsed by imagining himself on the sun and considering how the satellites would look from there. Clearly a satellite that could not be seen from the sun (because Jupiter stood in the way) would not be receiving light from the sun. So when a satellite is calculated first to be some distance east or west of Jupiter as seen from the earth, and is calculated second to be in the same line with Jupiter and the sun, it may be unseen because it is eclipsed by Jupiter. This second calculation does not even require additional labor, since it consists simply in leaving out the correction for parallax arising from motion of the earth.

Sagredo   Then before our friend discovered the need for that correction, he had in fact been calculating satellite positions for an observer on the sun.

Salviati   Say rather "at the center of Jupiter's orbit," from which the line along which the satellites are seen to move is always at right angles to the line of sight.

Sagredo   Very well; then, since very precise observations and calculations show that those two lines are not always at right angles for us observers on the earth, it is at least certain that the earth is not at the center of Jupiter's orbit. And because the same calculations enabled our friend successfully to predict actual satellite eclipses, which are deprivations of light from the sun, it followed that the sun must be quite near the center of Jupiter's orbit, if not exactly there.

Sarpi I have been listening with growing wonder and admiration. What you have been explaining to us, Salviati, is that the telescope revealed two proofs, and not just one, that the cosmology of Aristotle and the astronomy of Ptolemy are simply false and impossible. The phases of Venus proved that, and satellite eclipses confirmed it. Neither of those things could even have been known without the telescope. Is that right?

Salviati Yes, though one might say that our friend's earlier recognition of the need for parallax correction even to forecast satellite positions accurately had already shown that the earth is not at, or even near, the center of Jupiter's orbit. That that center is near the sun was suggested by the nature of the parallax correction, and any reason for doubt was removed by the subsequent detection of satellite eclipses.

Sagredo Then the telescope, by confuting Aristotle and Ptolemy, left open only the choice between Copernicus and Tycho. The phases of Venus, you said, were powerless to decide between those two systems. From what has been said thus far, the same seems to be true of Jupiter's satellites, which establish the sun as center of the planetary orbits, but not its rest or motion.

Salviati Very true; we must always distinguish in science what has been established from what is probable. Two things make the Copernican astronomy more probable than the Tychonic. The first of these is that our knowing the sun to be at the center of the orbit of Venus or of Jupiter supports the probability that there is a single center, and gives an indication that it is a center of motions for bodies distant from it, as the earth certainly is. The other is that in science, as in philosophy, the more probable view is that which makes the fewer assumptions.

Sarpi But was it not Tycho who made the fewer assumptions, by abandoning the Copernican hypothesis that the earth moves?

Salviati It would be, had Tycho not had to assume two motions of the sun not required by Copernicus. Both, moreover, must be shared by every planet, one motion daily at enormous speed and the other, in a different plane, annually. Although it is really hard to count and compare the assumptions, it is certainly simpler, with Copernicus, to eliminate enormous speed of the sun and of every planet by allowing the earth simply to rotate on its axis daily, and then to unify the apparent motions of the sun and planets by having the earth also travel annually around the zodiac. Besides, Tycho had to assume some power in the sun capable of carrying Saturn, Jupiter, and Mars millions of miles daily, but incapable of budging the earth.

D 269–71

Sarpi If the sun and the planets are not physical bodies like the earth, but are merely lights in the aethereal substance, that is no problem. Yet I see that Tycho's assumptions are not as few or as simple as I thought, so please continue. Excuse my digression.

Salviati Nothing is really a digression from our main purpose as long as it is related to the system of the world, as your question certainly is.

Going back now to our eclipses, the geographical problem of terrestrial longitudes is nothing more nor less than a problem of knowing the precise time of day at any place when it is a given time of day at some other agreed place, now usually taken as the Fortunate Isles. For most places hundreds or thousands of miles apart, that problem is most accurately solved by the timing of lunar eclipses. When our friend discovered the occurrence of satellite eclipses, he accordingly thought at once of their possible benefit to geographers, and especially to cartographers, above all in the mapping of colonies and territories in the New World. Not only are satellite eclipses much more frequent than lunar eclipses, but it is also easier to determine the exact time of their occurrences, those being more abrupt than is the entrance of the moon into the shadow of the earth, or its moment of complete reemergence from shadow.

GAW 258

On the heels of this idea came another, even more important, that could aid in transoceanic navigation. Except for about one month every year, Jupiter and its companions can be observed on

GAW 259   any clear night. Longitude of a ship at sea could therefore be approximated by comparing any observed position of the Medicean stars with tables of predicted positions at nightly times as seen from an agreed place, such as Florence. Perceiving the value of these ideas to countries having possessions in America, our friend

GAW 193, 257–59   at once began negotiations with Spain for exclusive use of tables he would prepare for use by naval officers whom he would train to make the observations at sea. For that reason he has kept secret the information I have just given you, which I ask that you likewise treat as confidential during those negotiations.

Sagredo   That is a very small price for us to pay for such a wonderful story of discovery and invention. Is this the reason for your present voyage to Spain?

Salviati   It is not the primary reason, though our friend thoroughly informed me so that I shall be able to explain anything at Madrid that is not fully understood from the long letters he has sent to the Tuscan ambassador there, for use in negotiations with the king and his advisers. That is how I happened to have at hand answers to many questions concerning the observations recorded in his book.

Sagredo   It was indeed surprising to me that you had ready so many explanations of curiosities and apparent anomalies in the text, and even spoke of printer's errors. I take it that you yourself have made calculations for comparison with past observations, using our friend's tables of the motions.

Salviati   I have indeed, and in that way I came to realize personally that in science there is no substitute for actual observation and calculation. Except for that brief section of Sizzi's book which we did consider, all the attacks against the *Starry Messenger* were made by opponents on the basis of few observations and no calculations at all. For that reason their authors may be said never entirely to grasp our friend's conception of science. Instead, they busy themselves with rejecting views that are philosophical rather than scientific, and for the most part are no less rejected by our friend.

Sarpi   You imply that they ascribe to him philosophical views he does not hold, or at least does not confuse with science as he sees it, though

they suppose that he must do so. In that case his best way of replying to them would be to write his promised book on the system of
the world, in which he could make clear the philosophical principles on which he did base his conclusions.

Sagredo   Salviati told Simplicio and me, when we were discussing the book
CES 10   on bodies in water, that the only conclusions our friend regards as
GAW 226   belonging to science are those based on sensate experiences and
             necessary demonstrations. Since observation and mathematics are
             not philosophical principles, I wonder whether he may not have
             given up philosophical principles as a basis for science.

Sarpi   That is hardly possible, even if anyone could have an incentive to
             make the attempt. We have already mentioned some assumptions
             made by our friend that are neither directly sensible nor necessarily
             demonstrated. For example, he assumed (as did Sizzi) that the
             motions of Jupiter's satellites must be regular, and must take place
             in circular orbits concentric with the planet. Such assumptions are
             tantamount to philosophical principles.

Salviati   Not necessarily, Fra Paolo. They may have been for Sizzi but not
             for our friend. What we read in the *Starry Messenger*, and what we
             have been talking about today—that is, the astronomical work our
             friend has done since—involved much observation, calculation,
             and reasoning, but little if any deduction from "philosophical
             principles" as that phrase is ordinarily used. Sagredo's remark and
             your answer to it are both worthy of consideration, but in my way
             of looking at things, as I said earlier, they concern propriety of
             terminology rather than matters of substance. I think that what-
             ever our friend writes on the system of the world, astute philoso-
             phers will oppose it—and not just a few, but the overwhelming
             majority. For them to say that it must nevertheless be based on
             philosophical principles cannot impose on our friend any obligation
             to identify such principles in his book; rather, it imposes on
             philosophers the task of identifying our friend's principles and
             exposing their falsity, which must itself be philosophical.

Sarpi   Or accepting them as sound and adopting our friend's system of
             the world. In either event our friend cannot be logically compelled

	to set forth philosophical principles unless he deduces his conclusions from them, and in what we have read and heard in our sessions there is no evidence of his having done that.
Sagredo	In his book on bodies in water he weighed the arguments of Aristotle and Democritus on a certain matter, finding the principles of both to have been defective. The conclusions of Archi-
CES 65	medes, he said, were correct because they were closely borne out by experience. A certain philosophical principle ascribed to Archimedes by a modern Peripatetic, he noted, was to be found nowhere
CES 62	asserted or implied by the Syracusan; and indeed our friend said that if it were to be found there, it would incline him to repudiate Archimedes. I mention this because it tends to support my idea, challenged by you, that our friend may have abandoned philosophical principles as a basis for science. I now add that, reversing the usual practice, he looks instead to science as a *basis* for philosophical principles.
Sarpi	That sounds to me a good deal like attempting to lift oneself by one's bootstraps.
Salviati	It seems so to you, Fra Paolo, because of the terminological debate that divides you and Sagredo. Our friend's present view is that
D 37–38	scientific investigations will enable us to philosophize better, and I have heard him say this on many occasions. In order for that to happen, the investigations must be free—or as free as possible—from philosophical preconceptions.
Sarpi	I have noticed that Copernicans often cite the saying of Albinus
CES 21	that in philosophizing the mind must be free.
Salviati	There is good reason for their doing so, Fra Paolo. One thing that cannot in any way be reconciled with accepted principles of natural philosophy is motion of the earth about the sun. That assumption is necessary in the Copernican astronomy, and it can never be accepted by Peripatetics. Since Aristotle's principles govern science in every university today, Copernicans have no hope of victory. They have little hope of getting even a fair hearing until it is recognized that for progress in philosophy the mind should be free, and not fettered as at present by Aristotle's principles of science.

Sarpi   It is clear that Aristotle rested his physics on immobility of the earth, since without that all his teachings about natural places and natural motions would become devoid of meaning. And now that I think of it, Plato also said quite clearly in the *Phaedo* that the earth is at the center of the heavens, is kept from falling or tending in any direction by the uniformity of the surrounding heavens, and will always remain in the same state and not deviate. Only the ancient Pythagoreans, called by Aristotle "the Italian philosophers," appear to have approved a moving earth. The Italian philosophers of today certainly do not do so; hence I see why Copernicans could prosper here only if astronomy were freed from bondage to philosophy.

Salviati   It is true that Copernicans would be no better off under Platonists
GAW 222   than under Peripatetics, as proved by a recent event. A Platonist professor at the University of Pisa has just denounced our friend's belief in motion of the earth as heresy, making that charge to the Grand Duchess of Tuscany when our friend was not present to reply.

Sarpi   Dear me; heresy is not in the jurisdiction of philosophers of any school, but solely in that of theologians. They may have discussed the question, though if any action had been taken on it I am sure I should know of it. Doubtless the professor was mistaken, and the saying of Cardinal Baronius that I mentioned earlier still prevails among responsible officials of the Church.

Salviati   A Benedictine abbot who was present spoke as a theologian in opposition to the philosopher's charge. You may know him, for he studied with our friend at Padua; his name is Benedetto Castelli, and he is now professor of mathematics at Pisa. But returning to the subject, our friend would be no happier if science were governed by the principles of the ancient Pythagoreans, whom you mentioned as tolerant of a moving earth. Freedom of science from philosophical principles means more than freedom from Peripatetic dogmas or Platonist schemes.

Sarpi   I thought the Pythagoreans were mathematicians above all, whose
D 11   principles science could safely adopt.

Salviati	Pythagoreans became fascinated by numbers to the point of mysticism that sometimes grips otherwise thoughtful men. You
iii, 212–14	remember that Sizzi, who later found an error in our friend's book, began by arguing a scientific question from occult properties of the number seven.
Sagredo	And so speciously that, as you know, I read no further and learned only from you that later on in his book he had raised a serious question based on our friend's observations. It is easy to guess what our friend thought of his arguments that the planets must be seven in number because holy candelabra have seven branches, and seven is the number of our eyes, ears, nostrils, and mouth, of the days in the week, and of the metals assigned to planets.
Salviati	Yes, and perhaps no less easy to guess is what he thought of Kepler's having reasoned in favor of Copernicus from the fact that in his system the planets are six, because there are just five Platonic solids.
Sarpi	This is very amusing; before we go on, I should like to be more fully informed about it. How on earth, or I should say how in heaven, does five prove six rather than seven to be the number of planets? I have never given much thought to hidden properties of numbers, but I should think that those who take them as philosophical principles would at least agree on which numbers govern the universe.
Salviati	Rather, such people are inclined first to pronounce on the system of the world and then to find properties of numbers that explain why it must be as it is. In the Aristotelian cosmology and the Ptolemaic astronomy there are seven planets, including sun and moon, circling the earth, which is supposed to be fixed at the center. Sizzi's mysticism accordingly was focused on the number seven, which has indeed attracted many such thinkers. In the Copernican astronomy only six planets circle the sun, because the sun is fixed, while the moon remains uniquely associated with the earth and is not counted as a planet. Kepler's mysticism accordingly focused on the number six, and was richly rewarded.
Sarpi	I can see that the number five is very special to mathematicians as

the number of regular solids, called "Platonic" and proved by Euclid to be exactly five. But I fail to see how that proves there are six planets and no more, dooming Ptolemy and Aristotle. I wish to learn, if Sagredo does not object to this digression.

Sagredo Not at all, though I already know how Kepler proceeded. Early in
GAW 41 our acquaintance, our friend received Kepler's *Prodromus to the Cosmographical Mystery* and discussed it with me, later putting it to use in a way that is not without interest in regard to the development of his own thoughts about the system of the world.

Salviati That is indeed news to me, so after I have replied to Fra Paolo, perhaps you will instruct us both. Kepler's reasoning, Fra Paolo, was purely mathematical, once he had decided for Copernicus on other grounds. Around each of the five Platonic solids it is possible to circumscribe a sphere. When all five are arranged in a certain order, starting from the sun, it is found that by circumscribing alternately spheres and regular solids, the relative distances of all six spheres are very nearly those of the six planets measured from the sun. That done, it is possible to find reasons why that order of the five Platonic solids, and no other, is appropriate to the six particular planets; and to cap his proof, Kepler deduced from this arrangement the Platonic harmony of the celestial spheres.

Sarpi Well, that is certainly one way to base science on philosophical principles, and I can easily see why our friend mistrusts an approach that led Sizzi to seven planets and Kepler to six. That makes me even more curious than Salviati to hear from Sagredo how Kepler's first book could have been put to use by our friend and has even influenced his thoughts on the system of the world.

Sagredo First I must say that although Salviati accurately answered your original question, he did not do full justice to Kepler. The great German astronomer is given to far-reaching mystical speculations, but he is by no means neglectful of very accurate measurements and painstaking calculations from them. That is now evident from his *New Astronomy Causally Considered*, published four years ago, but it was already seen in his first book, from which our friend took Kepler's Copernican figures as the most precise yet printed.

The first twenty chapters, devoted to the scheme described by Salviati, did not interest him, but what immediately followed them did. Kepler had next sought ratios connecting the observed motions of the planets with their mean distances from the sun. Kepler's analysis in terms of forces and souls did not appeal to our friend, so he applied to Kepler's astronomical measurements his

CES 29   own concept of *moment* and the principles of mechanics that connect distances with speeds. The result was quite remarkable,

GAW 63–65   and shortly before he left Padua he was able to connect that with his law governing the natural motions of heavy bodies here on earth, already mentioned by Fra Paolo.

Salviati   Sagredo, you incline me to mysticism, though not in physics and astronomy. I had no inkling of what you have just told us, our friend having never mentioned to me anything of the kind. It is as if some hidden power were guiding our discussions, when I thought I was directing them myself with a view only to providing a background against which we could best understand our friend's delays in publishing on the system of the world. At the outset I had the advantage of recent association with him, from which I knew that his project was much grander than one might suppose; I little dreamed that each of you, here in Venice, might possess information linking it to his work here, unknown to me. But let us defer further discussion of this until tomorrow, and take up now the attacks published against the *Starry Messenger* that gave our friend reason to delay his book on the system of the world, even beyond the time consumed in calculations and further discoveries of which we have already spoken.

Sagredo   An excellent suggestion. In preparation for this I placed on this table some books from my library that I thought you might wish to refer to. Two are included that were not attacks. One of them was misrepresented in Italy as having rejected our friend's *Starry Messenger*; the other corrected that misrepresentation and strongly defended our friend against his first and most vigorous opponent.

Salviati   Let me see what we have here. Ah, here is that first published attack, or rather that insolent libel which added innuendo to iii, 133–45 misrepresentation and relied on support from traditional authority against innovation in place of sober argument. It was published at Bologna by Martin Horky, a young Bohemian astronomer who had recently come to Italy to study under and to assist Professor Giovanni Antonio Magini, whose part in this deplorable affair is at least questionable. Though he protested his friendship toward our friend in letters, and expelled Horky from his house, he did not use his great reputation to lend support in any way to the new discoveries, at a time when that would have been of much help to our friend.

Sarpi I remember that when the governors of the University of Padua were debating the choice of a successor to the illustrious Giuseppe Moletti, long professor of mathematics to the honor of our university, this Magini was a strong candidate. He was a Paduan by birth, and though he held the chair of mathematical astronomy at Bologna, which he had won in competition with our friend, he was willing to relinquish that to return to his native city. I believe that he never fully recovered from his chagrin that the chair of mathematics at Padua was eventually given to our friend, after its having been left vacant for three years during the search for the best possible occupant.

Sagredo
GS 130–31 My own dealings with Professor Magini ten or more years ago left me doubtful of his good faith. They were in connection with an emissary of Tycho Brahe's who, after visiting Padua and Bologna, became most vituperative about me and our friend. Yet I had tried to assist him in his mission and sent to Magini my copy of Kepler's first book, though Magini later said he first knew of that from Horky several years afterward.

Salviati
iii, 129–40 Horky's book is called *Brief Excursion against the Starry Messenger*, and it is divided into four "problems" that we can mention later
iii, 202–50 The next attack to appear was Sizzi's *Dianoia Astronomical, Optical, and Physical*, published at Venice. It would not have been printed
iii, 208 had Horky not misrepresented Sizzi's position in a manner that was offensive to the young Florentine patrician. The *Dianoia* stressed supposedly scientific reasons for distrusting any observations not

made directly by eye, but only through optical instruments. I say "supposedly" because we shall see that Sizzi's objections were mainly of a philosophical character rather than securely grounded in the science of optics. I refer, of course, to arguments other than those we have already discussed, such as those he based on properties of numbers.

Sarpi   But, as we have already seen, that approach was also regarded as legitimately scientific by Kepler himself.

Salviati   I did not mean to deny that but to distinguish a kind of confusion of scientific with philosophical objections that is widespread and will have to be dealt with by our friend in his promised book. The third kind of attack is represented by this book, also printed here

iii, 311–93   in Venice, *On Phenomena in the Lunar Orb*, by Julius Caesar La Galla, a professor of philosophy at Rome. It is based entirely on

iii, 393–99   philosophical principles, concerning which our friend wrote some interesting notes as he read it in 1612.

Sagredo   Before you take these up, in whatever order and whatever detail you choose, you might say something of the other two books, which were not attacks against the new discoveries.

Salviati   I intended to do that, because both of them bear on any discussion of Horky's book, the first attack to appear in print. Even before that, support came in Kepler's *Conversation with the Starry Messen-*

iii, 99–125   *ger*, written in April 1610 (only a month after the new discoveries were announced) and published in May at Prague, where Kepler is court astronomer to the Holy Roman emperor. Far from being an

GS 134–38   attack, it was written as a letter to the Tuscan ambassador there at his personal request, as suggested by our friend himself. Although he was overjoyed to have such powerful support, rumors soon spread in Italy that Kepler had *refuted* our friend's new discoveries as well as his invention of the instrument with which they were made.

Sarpi   You were not here at the time, Sagredo, and the events that made possible so strange a reversal of Kepler's opinion are now generally forgotten. Horky's book alone was responsible, for his libel circulated in Italy before many copies of Kepler's *Conversation* had migrated here from Prague to contradict him. Anyone who read Kepler's book could see that he was at first incredulous and then became delighted and enthusiastic, not only accepting the Medicean stars but urging our friend to search for satellites of other planets, predicted to exist by Kepler on some mysteries of numbers similar to those which he had applied to Copernicanism long before.

Sagredo You mean that Horky read this acceptance of our friend's work and then deliberately published to the contrary?

Sarpi So disgraceful was Horky's entire conduct that I could say that that was the least of his offences. My friend Martin Hasdale at Prague,

x, 342–43 who was in touch with Kepler, wrote me of a letter sent by Horky to Kepler on 27 April 1610 that is almost unbelievable in its dishonesty. Had it arrived two weeks earlier (as in fact it could not have done), I doubt that Kepler's *Conversation* would have been written, let alone published.

Salviati I know the letter you mean, of which a copy has come into our friend's possession. He also received a letter from Kepler after

x, 416–17 Horky's *Excursion* was published, pointing out that Horky had completely misrepresented statements in the *Conversation* and authorizing our friend to publish the letter if he wanted to. But this other book, by a pupil of his, had made that unnecessary.

Sagredo I am curious to know about something else you said earlier. Our friend said clearly in the *Starry Messenger* that the invention of the telescope belonged to a Fleming, after which his own instrument was constructed. So when you said that rumors circulated in Italy that Kepler had refuted Galileo's claimed invention, you must have referred to something else, found by Horky in Kepler's book. What was it?

Salviati What Kepler said in his *Conversation* was this:

KC 15 So powerful a telescope seems an incredible undertaking to many persons, yet it [the invention] is neither impossible nor new. Nor was it [just] recently produced by the Dutch, but many years ago it was announced by Giovanni Battista della

Porta in his *Natural Magic*, book xvii, chapter 10, "The Effects of a Crystal Lens." And as evidence that even the combination of a concave with a convex lens is no novelty, let us quote Porta's words. Here is what he said: ". . . Through a concave lens you see distant objects small but clear; through a convex lens nearby objects larger, but blurred. If you know how to combine both types correctly, you will see far and near, larger and clearly."

Kepler then described other writings of Porta and some of his own, speculating whether the new invention had arisen from them or merely by accident. He then said:

KC 17   I do not advance these suggestions for the purpose of diminishing the glory of the practical inventor, whoever he was. . . . But here I am trying to induce the skeptical to have faith in your instrument.

Faith in the telescope is precisely what Horky tried to destroy. After him, in different ways, Sizzi and La Galla also discouraged belief in the instrument. Only Horky, however, pretended that Kepler had rejected our friend's discoveries. That impression remained alive in Italy until it was corrected by long citations from Kepler in this other book, called *Confutation of Four Problems*

iii, 149–99   *Proposed by Martin Horky against the Starry Messenger*, published by John Wedderburn at Padua shortly after our friend moved to Florence. We may speak of that when we have disposed of Horky's arguments.

Sagredo   I glanced at Horky's book when placing it here this morning and noticed something strange at once. Here on the first page is its opening "problem": "Whether there are four new planets around Jupiter." To make his answer conspicuous, he put in capital letters

iii, 137   toward the middle of the page the words NOVOS QUATUOR PLANETAS CIRCA IOVEM NON ESSE, and a bit farther down he wrote:

but since they do not exist in nature, I was never able to see the four new planets.

Yet in the second problem, "What are the new planets?" the author speaks of our friend's visit to Bologna in April 1610, during which he exhibited his telescope, thus:

iii, 140   on the night of 24 April I saw only two globes, or rather very tiny spots, and asked Galileo where the other two were hiding:
iii, 141   why did they not appear, the sky being clear? No response . . .
On 25 April . . . I saw with Galileo's telescope all four very tiny spots leaping from Jupiter, on the discovery of which he prides himself.

Sarpi   What seems strange to you, and to anyone of common sense, is that Horky seems first to deny in print the existence of something that he then goes on to admit he himself saw. This is a stale old trick of sophists and rhetoricians, by which they hope to induce others to conclude that the author under attack has simply lied and to read no further. Horky could not deny that something was seen, but only that the things seen were wandering stars around Jupiter. That is easy to say, and an honest opponent would say it at once, going on to give his reasons for interpreting the sight differently. Had the book been in Italian, we might suppose the foreigner to be guilty only of awkward phrasing of his "first problem," but there is no excuse for such paltering in Latin.

Sagredo   Rather, it makes things worse, since readers of Latin books are mainly scholars who would be even more offended than ordinary readers by such an insult to their intelligence.

Salviati   Think then how Kepler must have been insulted by that letter of which Fra Paolo spoke, sent by Horky after Galileo's visit, and therefore after Kepler had already expressed his confidence in the
x, 342–43   telescope and in our friend's new discoveries. The letter began by maligning our friend's appearance: hair falling out, skin blemished, brain doting, bloodshot eyes to observe seconds of arc, hands rheumatic from stealing philosophical and mathematical wealth, palpitating heart, physique unsuited to studious and great men, gouty feet. Next, the telescope was granted to work well for terrestrial observations but declared completely untrustworthy for celestial viewing, all at Bologna having agreed on this, Horky

said. According to his account, poor Galileo left Bologna defeated after two days, but not before Horky had taken wax impressions of the lenses, unknown to anyone, and promised soon to make a better telescope. Such was Horky's letter in reply to Kepler's inquiry.

Sagredo   It was foolish to attempt to impose on Kepler in such a way. No wonder he ordered Horky out of his house, as Magini had done at Bologna. But I disagree with Fra Paolo that if such a letter could have reached Kepler in time, he would not have written his *Conversation*.

Sarpi   Why, Sagredo? Horky's letter, as I recall it, was designed to leave Kepler with the impression that though many persons assembled at Bologna and tried for two nights to see what our friend described as new planets around Jupiter, not one person actually succeeded in perceiving any such thing. Had Kepler received that letter in time, he would hardly have risked his own reputation by writing in support of our friend's discoveries.

Sagredo   But that letter itself, being dishonestly written, betrayed the perfidy of the writer to so upright and intelligent a reader as Kepler. It is not necessary to dwell on the fact that the personal appearance of a man, even if revolting, could have no possible bearing on the truth or falsity of his observations and his interpretation of them. Unreliability of the letter would have been apparent to Kepler on quite other grounds, without this evidence of malice. As the foremost living writer on optics, Kepler would not for a moment believe that the same instrument could be trusted for terrestrial and be fallacious for celestial observations. And why, he would wonder, did Horky take wax impressions of the lenses if he believed the telescope untrustworthy? Once Kepler had read the *Starry Messenger*, he would certainly not have put any faith in Horky's dishonest letter.

Salviati   Nor is there much in Horky's book that need detain us. His arguments consisted mainly in showing that authority bore against the existence of any new planets. Tycho Brahe, the prince of observers among astronomers, had not seen any. Great astrologers had determined all the terrestrial influences of stars and planets without these pretended new ones. To this last, Wedderburn gave iii, 145 an amusing reply. Horky, in his fourth problem, "Why there are four Galilean planets in the sky," had asserted that new planets could serve only Galileo's fame, since they had no possible use in iii, 177–78 mathematical or optical theories. Wedderburn answered that they were in the sky to excruciate and perplex oversuperstitious people who strove to relate even tiny sparks in the heavens to particular effects. Wedderburn did much to correct the false impression of Kepler circulated by Horky, citing verbatim many long passages from the published *Conversation*.

Sagredo   I well remember this clever Scotsman who studied medicine at Padua and then went to Bohemia. He was a disciple of our friend who, in addition to defending him against Horky's malicious book, disclosed a use to which the telescope had been put that has iii, 164 not yet been revealed elsewhere. He had been present when our friend told Simplicio about some observations of insects at close range, magnified by the telescope. Horky having ridiculed our friend's statement that he had distinguished arcs as small as ten seconds in the sky, Wedderburn recounted observations of the eyes and other tiny organs of flies and moths. The structure of very small creatures is remarkably similar to that of animals better known to us.

Salviati   Such observations offer a powerful reply to a certain early common objection. To those who doubted that faith should be placed in things seen through the telescope alone, the only reply at first was that an object a mile away, as for example a church window, appeared through the telescope just as it appeared to the naked eye from a distance of a hundred paces or so; that is, twenty times closer. But at ordinary distances on earth the instrument had not previously revealed things entirely invisible to the unaided eye. Many people supposed that it could be trusted only for earthly, and not for celestial objects, because Aristotle had declared the two to be of entirely different natures and substances. What you have just told us would have provided an effective reply also to that, had our friend troubled to answer at all.

Sarpi *GAW* 286, 290	Because the study of anatomy was long my special hobby, I hope our friend will some day make an instrument less unwieldy than his telescope for investigations of small structures at close range, with advantages to surgeons and anatomists no less than to people curious about small plants and animals. To see the moon as if it were but one terrestrial diameter distant is glorious, but of less use to mankind than it would be to have eyes thirty times more powerful than ours for examining things more closely affecting our lives.
Salviati iii, 221–23	What comes into consideration here is a philosophical question related to any extension of sensate experiences. Let me pass on to Sizzi's book in order to explain what I mean. His main theme was that Alhazen and others expert in the science of optics had given many reasons for which simple vision differs from reflected, or refracted vision. Those reasons, he said, are imprudently ignored by supporters of our friend's claimed discoveries in the heavens, which are all necessarily made by refracted vision. The objection, founded on optical science, becomes one of philosophical principle when the reality of new things in the heavens is thus called into doubt; for if there is one thing that concerns philosophers above all, it is applicability of the concept of reality.
Sarpi	How right you are, and how easily the point escapes us when the philosophical question masquerades as one of science, as in this case. It certainly falls within the province of the science of optics to say that simple vision differs from refracted vision. But it is another matter to say that only what is seen by simple vision is real. That would make a fiction or an illusion of everything presented to us exclusively by refracted light. To assert that as if it were a scientific truth, rather than a philosophical opinion, is most misleading. Indeed, light is refracted even in our own eyes.
Salviati	I believe that Sizzi was himself misled and that, if he thereby misled others, it was not intentional on his part. Unlike Horky, Sizzi was in search of truth. Before we proceed to discuss the philosophical issue, if we decide to do that, let us examine the relation of Sizzi's book to Horky's. In his preface to the *Dianoia*, Sizzi spoke of our friend's visit to Pisa and Florence, where many persons were shown the telescope and its revelations. He wrote:

iii, 207   I also, on one occasion, took part in the observations, and I saw the same phenomenon as all others present, but in some particulars it did not appear to me as to them (and I shall say more of this in my book). That seen, and I having attentively considered the structure of the telescope, it quickly came to me that in such phenomena there was some mistake and deception of sight; so that after this pondering I set down in writing some things drawn from the mysteries of optics and astronomy.

Sizzi's reflections were not intended for publication and would not have been printed had it not been that Horky, who knew of them, named Sizzi in his book as someone who agreed with him. The well-bred Florentine was shocked to be associated with the malicious conduct of Horky and proceeded to publish his own work, in which he assured Galileo that his sole interest was in finding truth. At that time he believed that truth lay on the side of the great majority of philosophers. I told you yesterday how his views changed after he moved to Paris.

Sagredo   I take it, then, that in the part of his book I did not read, in addition to criticisms offered concerning inconsistencies among the observations published by our friend, there is a discussion of the science of optics and the theory of refracted vision, drawn from the most expert authors.

Salviati   Alas, no; the young Florentine was but a dilettante in science. In vindicating his character and intentions, I did not mean to exaggerate his achievements, which were not great even in the section of his book from which we read a small part yesterday. Instead of discussing refracted vision or explaining a theory of it, Sizzi explained problems on the authority of good writers to discredit the placing of any faith in the telescope. Thus, against those who supported the reality of the Medicean stars, he wrote that they failed to know:

iii, 220   that the viewing takes place through different diaphanous mediums denser than air, which are crystals and glasses, very dense and a bit concave, through a tube of tin or wood of a certain length and shape that draws and holds the visual property of a ray, and its angle, through foggy night air—they ignoring the best optical theorists, for whom, whenever we look at anything through a denser medium, there is always necessarily a refraction at the surface of the denser body.

And so on. Magnification, he said, like multiplication of images, takes place not through simple vision but by refraction. What is thus seen is no more real than are parhelia that seem to multiply the number of suns in the sky. Nor can reality be concluded from the fact that the starlets appear now to the east and again to the west of Jupiter, and:

iii, 221–22   likewise from the fact that we see now two, and now three, it is not legitimate to conclude real existence of stars turning about Jupiter.

Sarpi   If that is a fair sample, this writer took his clue from those philosophers, infinite in number and variety, who without being skeptics take joy in raising every possible reason to doubt anything newly proposed, as if everything that can be known has already been established. Young men particularly delight in that way of showing the vast extent, not of their knowledge, but of knowledge stored in books they have read without troubling to learn difficult things, but only the existence of difficulties. If I am correct, I should expect Sizzi to exhibit also a flair for logical minutiae.

Salviati   Not only are you right, but in displaying this he permits himself to be sarcastic about innovators and would-be discoverers. An example occurs when he rejects the common-sense conclusion from the fact that moving starlets are telescopically seen near Jupiter but not near other planets. That does not remove them from the realm of mere optical illusion, he says:

iii, 224   I do not see that the conclusion they deduce follows, for in the whole realm of logic it is not the case that the best conclusion can be inducted from dissimilar facts. I do not know whether in this recent astronomy they have introduced such a new form of concluding, and just as they seek to introduce new planets, so they have been able to introduce that new form of inference; but among logicians, the reasoning is not valid.

Sarpi   I sigh to hear this young man arguing as if the old astronomy had not likewise been "inducted from dissimilar facts." That is mere jargon, since all deduction is no less dependent on differences than on similarities. It is proper for science to treat sensate experiences as facts, and it is as much deduction as induction to say "Here I see something, but not there; hence something is here that is not there."

Salviati   So much for Sizzi's doubts about the reality of the Medicean stars. Although Wedderburn replied only to Horky, one general remark of his equally applies to Sizzi and should also be kept in mind when we deal with La Galla's book. Wedderburn's remark almost certainly reflects the thought of his teacher, our friend:

iii, 168–69   . . . their writings are prodigal of flourishes and phrases, and they do not hesitate to write whole pages when they wish to blame, find fault, or elude someone, but when it comes to the point at issue that they ought to open and argue and follow out, [they] are suddenly at a loss and troubled in drawing this from that or deducing that from the other. This certainly seems to me [to reflect] too little the behavior of the mathematician, or to imitate these disciplines in which nothing is undefined, and everything coheres. Such sciences alone deserved in antiquity, and by Aristotle in particular, the name of method; from them all other sciences derive their method (if they have a genuine method), and by resemblance thereto link together the forms, principles, and affects of the objects, each in its own rank. Hence it is that the entire philosophy of Aristotle is rife with many examples from geometry, where he says something comes up that is difficult and quite inexplicable in any other way. But the Philosopher seems to excise that part which is the principal object of a precise science, over the discovery of which more glory redounds than from the rest—so much so that logic, in so far as it is the instrument of knowledge, should be said to be the legitimate offspring of mathematics.

Sagredo   This seems to praise Aristotle for founding logic as his basis for science, but to blame him for failing to perceive that logic is born

of mathematics. I wonder that it omits praise of Plato, who put
mathematics alone as the instrument of science.

Salviati   In this place Wedderburn perhaps paraphrased some notes left
by our friend, who conceives science as the domain of sensate
experiences and necessary demonstrations. Plato would not have
given those two things equal rank and joint proprietorship of
science. Aristotle did, but he then excised mathematics from
natural philosophy. If the conception is Wedderburn's own, it
at any rate catches the spirit of our friend's present view of the
requirements of true science.

Sarpi   That view is by no means foreign to an opinion he held while he
GAW   was still here, known to me through the many discussions we had
100–103   of the science of natural motions of heavy bodies near the earth. In
GAW 86–90   the course of his studies of such motions he came to discover, about
ten years ago, that an exact mathematical rule connects the distances in descent from rest with the times of such motions. There
was no doubt that the rule was true, because careful measurements
of such motions repeatedly confirmed it. Yet, although Aristotle
placed natural motions at the very basis of physics, this mathematical rule remained unsuspected by countless philosophers who
had written commentaries on Aristotle's *Physics* and long treatises
on natural motion. We concluded that this neglect could be explained only by the fact that Aristotle had excluded from natural
philosophy the use of mathematical practices, as alien to the understanding of physical phenomena. As I said, that was ten years
ago, so I am not surprised that our friend's pupil, Wedderburn,
reproached the Philosopher for excising geometrical problems from
physics. Certainly our friend's present view of science is mathematical in spirit.

Sagredo   There is also another side to the neglect that Fra Paolo mentioned,
and the explanation that he and our friend found for it. If careful
measurements many times repeated always agree with our friend's
mathematical rule, anyone could have found that rule in the course
of making such measurements, long before our time. Yet the rule

remained unsuspected by innumerable philosophers who studied and wrote treatises on natural motions, so it is probable that they neglected sensate experiences as a principal source of scientific knowledge, though Aristotle himself by no means excluded those and even said that evidence of our senses can refute any conclusion reached by reasoning. I mention this, Salviati, because you have told us that precisely these two things—sensate experiences and necessary demonstrations, especially mathematical ones—now define the field of true science for our friend.

Salviati   By your perceptive comments, you and Fra Paolo have greatly simplified for me my remaining task of dealing with the last of these three books, which is very long, and, though it certainly deserves our attention, I should be sorry to have to examine the
iii, 323–25   arguments in detail. Fortunately for us Professor La Galla, being a philosopher, set forth in the second chapter, after the usual proem, the method to be employed in his entire disputation. Our friend wrote some comments on that chapter that I recall pretty well. By considering La Galla's philosophical basis of attack and our friend's remarks on it, we may be able to dispense with the author's verbose arguments in detail. My reason for saying this is that the methodical considerations listed by La Galla at the very outset relate precisely to the two defining criteria underlying our friend's science just named by Sagredo, both being rejected by the Roman professor. But since it is getting late, and in any case I should like a little time to refresh my memory about this book, I suggest that we defer this until tomorrow.

Sagredo   That seems to me an excellent idea, since our task tomorrow is to reach as full an understanding as we can about our friend's long delay in writing his promised book on the system of the world. No better beginning for that can be imagined than the conception of science prevailing among professors of natural philosophy, for whom La Galla may be considered an authoritative spokesman.

Salviati   Very well. But remember, Fra Paolo, that for our final session you must bring along your old notes on our friend's explanation of the tides, for if I am not mistaken that is going to contribute a crucial element to our speculations about the promised book. We have already several reasons to believe that it will deal not only with astronomy, but also with physics, as would be the case if in it our friend connects celestial motions with the motions of our seas.

Sagredo   Speaking of the seas, we should dine tonight in the restaurant most famed of all in Venice for its delicacies taken fresh from our sea and served in ways that delight the eye no less than the palate. You have not been there, Salviati, because it is situated on the Lido, which you must also see before you leave Venice. The journey by gondola will take us some little time, and as it is getting late we should start at once. Will you come with us, Fra Paolo?

Sarpi   Thank you, but Salviati has reminded me that I must find my old notes on the tide theory of our friend, which (as I remember them) were brief and lie buried in a great mass of papers written long ago. With your permission I shall return to my convent and start the search, as I should have done before. But I shall be here with them tomorrow, without fail.

Sagredo   Good night, then, and let us be on our way, Salviati.

Salviati   Good night, Fra Paolo. Your wish is my command, Sagredo.

End of the Third Day

THE FOURTH DAY

Sagredo   While we are waiting for Fra Paolo, you may be interested in something that occurred to me after you suggested that we limit our consideration of Professor La Galla's book to his method of attack on the *Starry Messenger*. Several years ago, about the time I left Venice to serve as treasurer in Palma, our friend told me his view of proper method in science, which might be appropriate this morning.

Salviati   Certainly it would interest me, if it bears on his work here relating to the science of motion. Except for his studies of bodies in water I know little of his method in physics, our conversations at Florence having been chiefly about astronomy.

Sagredo   It does bear on that, though curiously enough, the source of his method was in ancient astronomy rather than in previous physics.

Salviati   In that case it will be most appropriate for you to tell us about
it, for, as I have said more than once, it has begun to appear to
me that our friend's system of the world will combine astronomy
with physics, contrary to all past tradition, excepting Kepler
alone.

Sagredo   Good; and there, if I am not mistaken, comes Fra Paolo now,
hurrying toward us with a great volume under his arm.
Good morning, Fra Paolo.

Sarpi   Good morning, gentlemen; please excuse my tardiness.

Salviati   Good morning, and welcome. I see you have brought a great
book in place of the few notes we expected. Can it be that our
friend had that much to say about tides?

Sarpi   Heavens, no; there are but four short paragraphs here on that
subject. It is just as well that I did not accompany you last
night, which I spent leafing in vain through old accumulated papers looking for the notes I promised to bring. This morning I
realized that my only hope of keeping the promise would be to
find that I had copied them into this volume, in which I used to
enter various reflections that I might some day draw on for a
book about natural philosophy. Fortunately I had done so, and
only a few minutes ago I found them, preceded by more than five
hundred unrelated notes.[41] That is my reason for being a bit late,
and I am sure you will forgive me when I tell you that I see they
were written as long ago as 1595, not long after our friend began
to lecture at Padua.

Salviati   There is no need here of pardon, but only of thanks for your
trouble. Before we hear these notes, I was going to speak of the
method Professor La Galla employed against our friend's conclusions, and now Sagredo has just mentioned something else that
perhaps we should hear first.

Sarpi   What is that, Sagredo?

Sagredo   It concerns proper method in science when its objects are amena-
ble to precise observation and measurement.

Sarpi   A most fitting opening for this final discussion, in which we are to speculate about our friend's system of the world. Please let us hear it at once.

Sagredo In the wake of that great quarrel over the location of the new star
GAP 11–13 of 1604, our friend and I were talking about Professor Simplicio's disdain for measurement and his insistence on philosophical principles governing materials and qualities when deciding a question of cosmology, or rather of cosmography. Two or three years before, our friend had added to his lectures on cosmography for beginning students a little introduction to the method that astronomers have followed ever since Ptolemy. Let me get my copy and read this; it is quite short.

Sarpi While Sagredo is away, Salviati, I might explain that Simplicio argued in 1604 that methods of measurement used on earth could not safely be extended to the vast distances of heavenly bodies.

GAP 38 Our friend maintained that sighting and triangulation, as practiced by surveyors and confirmed by ordinary measurement, could be validly extended to any distances whatever. When applied to the new star they proved it to be far beyond the moon, contradicting the Aristotelian principle that no change can take place in the heavens.

But here comes Sagredo.

Sagredo It was to make astronomical science easily understood by beginning students that our friend began his treatise on the sphere as follows:

☆ ii, 211–12 In cosmography, as in the other sciences, the subject should be identified and then the order and method to be observed in it should be touched on. Accordingly we say that the subject of cosmography is the world, or we mean the Universe, as indicated to us by the word itself, which means no more than "description of the world." But of everything that might be considered about the world, only a part belongs to cosmography, and that is to reflect on the number of the parts of the world, their arrangement, and the shape, size, and distance thereof, and principally their motions, leaving consideration of the qualities of such parts to the natural philosopher.

As to method, the cosmographer customarily proceeds in his reflections in four steps. The first embraces the appearances or "phenomena," and these are nothing but sensate observations we see in daily life, such as the rising and setting of stars. . . . Second come the "hypotheses," which are no more than suppositions relating to the structure of celestial orbs so as to correspond with the appearances—as when, guided by what appears to us, we should assume the heavens to be spherical and to be moved circularly. . . . Third there follow geometrical demonstrations in which, by means of some properties of circles and straight lines, the particular events that follow from the hypotheses are demonstrated. And finally, what has been demonstrated from lines being calculated arithmetically, this is reduced to tables from which we may without trouble find the celestial arrangement at any moment of time.

Salviati	Admirably put for beginning pupils, and if I am not mistaken this is precisely the method established by Ptolemy in his *Almagest*, who explicitly left to physicists and to metaphysicians anything over and above this observational and mathematical science of astronomy.[42] But I believe you said, before Fra Paolo arrived, that our friend somehow related this method to physics, which would seem to defy Ptolemy.
Sagredo	Not exactly. Ptolemy wrote that agreement in physics could never be hoped for among philosophers because matter is changeable and inconstant. When our friend wrote his introduction, he had not yet tried to apply Ptolemy's method to motions here on earth. But shortly before the new star appeared, he told me, he had succeeded in measuring distances, times, and speeds in the natural descent of heavy bodies. During the debate over location of the new star, he argued astronomical measurement against Aristotelian principles. From that time on, our friend adopted in physics the method that had made astronomy an exact science, achieving the same certainty in his new science of natural motions.

Salviati   Let me see if I fully understand this hint of yours, Sagredo. Astronomy had never had anything to go on except measurements of times and angles. At the very basis of physics, according to Aristotle himself, lay local motions. Natural motions, carefully measured, could similarly yield a science built on appearances, suppositions, geometrical deductions, and arithmetical calculations. Was that the idea?

Sagredo   That was exactly the idea. The inconstancy of matter need not be a source of perennial disagreement about physics, if matter as such ceased to be considered—just as had been done in ancient astronomy on the advice of Geminus.

Sarpi   But can matter as such be disregarded in physics? Aristotle would never say so, and I am inclined to agree with him.

Sagredo   In the method of suppositions based on appearances, established by Ptolemy, we may disregard anything we wish to. To have a science of natural *motions* of heavy bodies we need not consider the *material* of such bodies, unless observations contradict our supposition that it can be disregarded. Before astronomers applied careful measurements, philosophers probably considered it impossible to ignore the celestial substance when investigating heavenly motions.

Sarpi   I see your point. Our friend's new science of motion would fall far short of a complete physics, if (as I believe) completeness requires consideration of matter. Yet it would be a valuable, even a fundamental part of physics, by which that might be linked to astronomy in our friend's system of the world. In that view, far from our friend's having defied Ptolemy, he took from him his clue to a new science.

Salviati   So it seems. Though I am surprised that so interesting a conception has not been mentioned to me by our friend, I am infinitely obliged to you, Sagredo, for having done so. But let us now turn to the method adopted by La Galla as a true Aristotelian. Here is
iii, 311–93   his book, *On Phenomena in the Lunar Orb*, and I turn to the second chapter, entitled "By What Method the Disputation Is to Be Conducted." Putting aside what we have just learned about our friend's conception of science, let us see what method the Peripatetic

professor offered, and how our friend replied in his notes. First there is this dictum:

iii, 323

> Geometry treats of the continuum and Arithmetic of discrete quantity; and though Mathematics sometimes stoops to physical quantities, as in Optics and Harmonics, nothing in those contemplations affects physical and natural qualities, which it [mathematics] considers only as accidents. So it is manifest that those disciplines are conducted only theoretically and scientifically, not practically and mechanically.

Sagredo The basis of La Galla's attack is so different from Sizzi's, which we heard yesterday, that it seems those two opponents of our friend were contradicting one another. For if mathematical optics has nothing to do with practice, it was wrong for Sizzi to argue from that science against actual telescopic observations.

Sarpi On the other hand we might say that the two adversaries have jointly proved our friend to be wrong, no matter whether mathematics is or is not partly physical; and since it must be one or the other, he is absolutely wrong.

Sagredo He might reply in defence that since two men who argue from philosophical principles cannot agree on even so clear a matter as the nature of mathematical considerations, both are incompetent to decide scientific questions that involve mathematical calculations.

Salviati What our friend wrote in the margin, as nearly as I can remember, was this:

☆

iii, 323–24

It was questioned whether mathematical considerations of the heavens are physical or mathematical: if they are in fact purely mathematical, then predicted times of eclipses, conjunctions, and so on do not agree with those sensible and real conjunctions.

It is no less ridiculous to say that the geometrical does not correspond with the material than if one were to say that arithmetical properties in sensible bodies do not correspond to numbers—that, for instance, rules for ordering and coordinating armies do not work when we deal with military bodies.

Sagredo	This is amusing. In effect our friend says: First tell me whether astronomers predict eclipses exactly, or only approximately? If exactly, there is no way to distinguish the purely mathematical from the physical in astronomy, against your thesis. But if the agreement is approximate, then that immeasurable gulf you create between theory and practice becomes measurable, and may be no wider than a gnat's eyebrow. Only by making calculations and observations would it become possible for La Galla to answer the question; and if he made them, even he would see how foolish was the thesis that had inspired his doubt.
Sarpi	Sagredo, you make our friend's conception of science resemble a judge's conception of law reflected in the maxim *de minimis non curat lex*.
Salviati	And that is true, for I remember that in another note he wrote:

☆ iii, 394   It is further necessary to see whether the sensibles that you say deceive astronomers are minimal, or maximal, etc.

Sarpi	In justice it would be necessary for philosophers who raise these objections to make such determinations, and I am not sure that they are able to observe and calculate. Astronomers, on the other hand, continually make the determinations by employing procedures that La Galla implies to be untrustworthy as the basis of science. The history of astronomy has been a history of reducing discrepancies between prediction and observation by continually improving observations and modifying calculations. Without that process no science of astronomy would have been possible, but only cosmological speculation.
Sagredo	If I understand La Galla correctly, he means to say that a science of astronomy in Aristotle's sense *is* impossible and that what we call astronomy is not a science but merely a practice.[43] The Aristotelian science of cosmology was developed in *De caelo* by pure reason, so it remains incapable of alteration by mere observation—either for

better or for worse. Philosophers do not feel obliged to make any determination about the magnitudes of any sensibles that deceive astronomers; it suffices for them to show that astronomers are deceived. Since no true science can be built on deceptions, whether maximal or minimal, the pretended science of astronomy is effortlessly destroyed by philosophers with hardly more than the stroke of a pen. All that is necessary is to show that the sense of sight deceives us, as I suppose La Galla next tried to do.

Salviati   That is correct; most of the rest of his chapter on the method by which he would impugn our friend's claimed discoveries was devoted to the deceptions revealed by optics, or perspective as it is

iii, 325   commonly called. Thus he says that a picture painted on a flat surface may appear to exhibit roughness, prominences, and depressions that do not exist at all on such a surface. As I recall it, our friend asked:

☆ iii, 394   If these deceptions are made by perspective, who can better correct and understand them than those same perspectivists?

By this term he meant, of course, not painters but mathematicians who study optics, or perspective. Elsewhere in his notes he amplified his view:

☆ iii, 397   You want to charge mathematicians with ignorance for not perceiving that sense deceives us in commonly perceived things, as if whether knowledge of them is or is not deceived were a recondite and most profound mystery and secret of philosophy. But who have made more numerous and more exact observations and theories about the deception of vision than those same mathematicians?

Sagredo   It seems that La Galla was so busy condemning mathematicians for doing mathematics as to forget what else they do, at least those of them who were not born blind. For the purposes of pure mathematics, all that are needed are definitions and axioms. Yet in fact many mathematicians also make observations, and in that way optics was born, and astronomy, and even harmonics if we believe the story about Pythagoras and the blacksmith.[44] For want of a better word those are also called *sciences*. Though philosophers might deny to them that honorable name, they would have been unable to create them without mathematics, and I do not see how they are entitled to recognition as experts about studies they could not have created.

Sarpi They might be regarded as experts if they could at least improve those studies, but La Galla seems not to offer to do that. Rather, he seems to caution men not to believe their own eyes without first consulting philosophers, who, he has already said, divide the theoretical and scientific from the practical. By that means, I take it, he will prove from the science of optics that the practice of astronomy is fallacious.

Salviati So it appeared to our friend, who remarked in his notes:

☆ iii, 395 From the Author's statement, if men had been born blind philosophy would be more perfect, because it would lack many false assumptions that have been taken from the sense of sight.

With that, I believe, the gulf that separates our friend's view of science from that of philosophers is evident. And now I should like to hear from you, Fra Paolo, that explanation of the tides by which our friend connected motions of the seas with motions of the earth itself. Finally we shall both learn from you, Sagredo, what connection our friend saw between the speeds of the planets in their orbits and the motions of heavy bodies here on earth.

Sarpi I see what you are driving at. You have said several times that there may be much more to our friend's system of the world than a choice between the various rival hypotheses used by astronomers. If I am correct, you mean that when our friend's promised book does D 9, 15–19 appear, it will challenge the basic principle by which Aristotle sharply distinguished celestial phenomena from elemental motions by his drawing a barrier at the orb of the moon to divide natural motions into the straight and the circular. That would mean that along with arguments for the new astronomy, our friend will present as part of his system his science of natural motions of heavy bodies here on earth.

D 22–28, 138–230

Salviati   So it seems to me, as was hinted in his second mention of the promised book in his *Starry Messenger*. During the past three years at Florence he has been too occupied with astronomy to have said much about terrestrial physics, so I should not have guessed at his further intention had it not been for your remarks, and Sagredo's, about his investigations here before he had the telescope. Even now I have but caught a glimpse of those and must ask for further instruction.

Sagredo   It is only fair that just as you have told us what he has done in astronomy since he left Padua, we should tell you what we know of his work on motion while here. In that way we shall all be equally informed when we guess at reasons for continued delay of his book on the system of the world that gave rise to our discussions. As a beginning, I wish to read some passages from his recent *Sunspot Letters* that I believe will throw light on the scope of the promised book as he now views it. The first is this, after the motion of sunspots had been ascribed to their lying on the very face of the sun and being carried around by its monthly rotation, never before suspected. Replying to an alternative possibility, our friend wrote:

☆

D&O 113–14

But if anyone should wish to have the rotation of the spots around the sun proceed from motion that resides in the ambient [aether] and not in the sun, I think it would be necessary in any case for that ambient to communicate its movement to the solar globe as well. For I seem to have observed that physical bodies have natural inclination to some motion (as heavy bodies downward), which motion is exercised by them through an intrinsic property and without need of a particular external mover, whenever they are not impeded by some obstacle. And to some other motion they have a repugnance (as the same heavy bodies to motion upward), and therefore they never move in that manner unless thrown violently by an external mover. Finally, to some movements they are indifferent, as are those same heavy bodies to horizontal motion, to which they have neither inclination (since it is not toward the center of the earth) nor repugnance (since it does not carry them away from that center).

And therefore, all external impediments removed, a heavy body on a spherical surface concentric with the earth will be indifferent to rest and to movements toward any part of the horizon. And the body will maintain itself in that state in which it has been placed; that is, if placed in a state of rest, it will conserve that; and if placed in movement toward the west, for example, it will maintain itself in that movement. Thus a ship, for instance, having once received some impetus through the tranquil sea, would move continually around our globe without ever stopping, and placed at rest, it would perpetually remain at rest, if in the first case all extrinsic impediments [to motion] could be removed, and in the second case no external cause of motion were added.

Salviati   From this I see how our friend may be able in his promised book to remove several physical objections to motion of the earth. I remember, Sagredo, that in our conversations with Simplicio about

CES 121–22 bodies in water you said that you had tried the experiment of moving a heavy beam in still water by pulling it with a slender string and were surprised that with patience that could not only be done, but that when the string broke the beam continued at the same speed for a long time.

Sarpi   Aristotelians will strongly contend that everything moved must be moved by something else, no body being able to move itself.

Salviati   Very true, but misunderstood by them, as by Aristotle, to mean that something in *contact* with the moved body must be found at every moment during the motion. The beam is moved by the string, but then maintains itself in motion to which it has no repugnance, and uniformly if it has no natural tendency to that motion. Excuse me for this interruption, Sagredo, but seeing a clue here to the new science of motion worthy of further thought, I wished to call attention to it at once. Please continue.

Sagredo Having founded this reflection on observation of heavy bodies in progressive motion, our friend went on to apply it to simple rotation:

☆ *D&O* 114 Now if this is true (as indeed it is), what would a body of ambiguous nature do if continually surrounded by an ambient that moved with a motion to which the body was indifferent? I do not see how one can doubt that it would move with the motion of the ambient. And the sun, a body of spherical shape and balanced on its own center, cannot fail to follow the motion of its ambient, having no intrinsic repugnance or extrinsic impediment to rotation. It cannot have an internal repugnance, because by such a rotation it is neither removed from its place, nor are its parts permuted among themselves. Their natural arrangement is not changed in any way, so that as far as the constitution of its parts is concerned, such movement is as if it did not exist. . . . It does not appear that any movable body can have a repugnance to a movement without its having a natural propensity to the opposite motion, for in indifference no repugnance exists; hence anyone who wants to give the sun a resistance to the circular motion of its ambient would be putting in it a natural propensity for circular motion opposite to that. But this cannot appeal to any balanced mind.

Salviati I am very glad that you recalled and read this, as it shows a way in which our friend's system of the world will combine physics with astronomy, and further advances my knowledge of his progress toward a science of natural motions. Was there something else you wished to read?

Sagredo The other thing I had in mind was this, bearing on the relation of natural science to philosophy:

☆ *D&O* 123   In our speculating we either seek to penetrate the true and internal essences of natural substances, or content ourselves with knowledge of some of their properties. The former I hold to be as impossible an undertaking with regard to the closest elemental substances as with more remote celestial things.

Sarpi   Excuse me for interrupting, but is not Aristotle's goal in natural philosophy rejected here as impossible?

Sagredo   Impossible of attainment in science as our friend conceives it, and therefore left to metaphysics, as is seen a few sentences later:

☆ D&O 124   I know no more about the true essences of earth or fire than about those of the moon or sun, for that knowledge is withheld from us until we reach the state of blessedness. But if what we wish to fix in our minds is the apprehension of some properties of things, then it seems to me that we need not despair of our ability to acquire this respecting distant bodies just as well as those close at hand—and perhaps in some instances even more precisely in the former than in the latter. Who does not understand the periods and movements of the planets better than those of the waters in our various oceans?

Salviati Please stop here, Sagredo, if you were going to continue. Enough has been read to show that our friend now takes the method of science to be that of inquiring into properties of things, leaving their inner natures to metaphysicians. It is fortunate that he selected the motions of planets and the tides as his first example, for that reminded me that we have invited Fra Paolo to tell us how our friend long ago proposed to explain the tides.

Sarpi   As I said, these notes are in a journal I once kept for jotting down thoughts on natural philosophy, put aside when I was drawn into the service of our Most Serene Republic. It is a miscellaneous collection of observations, reflections, and things heard from others, which I once intended to put in order and perhaps to refine and publish. Before reading the four consecutive paragraphs about tides, I shall explain how they came to be here. When he lived at Padua, our friend often visited Venice, usually crossing from the mainland on a barge carrying fresh water to our city. On one such occasion he came to see me in some excitement, saying that he had just hit upon an explanation of the daily rising and falling of the seas, if one could believe with Copernicus that the earth itself moves with two different circular motions. [45]

Sagredo I wonder if he had discussed that astronomical hypothesis with you before this time, which you said was in 1595.

Sarpi I believe not. In those days our common interest was in problems of mechanics and of motion, as I remember. Why do you ask?

Sagredo I was barely acquainted with him in 1595, but I recall that about
D 128 that time some lectures were given at Padua on the Copernican astronomy by a foreigner called Wursteisen [46] who had studied at the University of Rostock. I did not attend his lectures, because in those days the idea of moving the earth seemed to me a majestic folly. Chancing not long afterward to meet with our friend, who seemed to me a thoughtful if rather conservative mathematician, I mentioned that opinion to him. To my surprise he replied that motions of the earth were not impossible and that one should not be so wedded to any belief as to be unwilling to hear arguments to the contrary. In this matter, he said, there were grave difficulties for both parties. But I do not recall that he talked much about astronomy until that controversy over the new star of 1604.

Sarpi Neither do I. Possibly his first preference for the Copernican view arose from this speculation about the tides that I am about to read. When we first began talking about problems of motion that fascinated us both, he showed me a treatise *De motu* that he had
M&M 15 written not long before, at Pisa. Though it contained many novel ideas, I remember that the earth was treated in it as motionless at the center of the universe. Hence it is clear that he had found no reason to prefer the Copernican system before he came to Padua at the end of 1592.

Sagredo I also saw that treatise and remember that the daily rotation of
M&M 74 the fixed stars was treated as a real motion, not as an appearance created by rotation of the earth. What can have altered his opinion after he came here?

Sarpi   That is what I was coming to. Our friend certainly did not remain
ignorant of the Copernican system while at Pisa, because in that
same treatise *De motu* he mentioned a certain ingenious Copernican

M&M 97   mathematical theorem in *De revolutionibus* by which straight motion was derived from a combination of circular motions. That also
confirms his early interest in motion for its own sake rather than in
astronomy as such. Now, there is a way in which continual study of
natural motions of heavy bodies could give rise to interest in the
new astronomy, whereas I do not see how the latter could as easily
give rise to the former. That tends to confirm my opinion that our
friend's original preference for the Copernican astronomy grew out
of certain natural motions of heavy bodies on earth. Perhaps for a
long time his preference lacked any other tangible support, which
was finally supplied by the new telescopic discoveries.

Salviati   Whether or not you are right, the idea is interesting, and so
unexpected that this rather long digression has been more than
justified. I suspect that your "certain natural motions of heavy
bodies" will turn out to be the tides, considering the context of
your digression. But it would be foolish for me to pretend that
I have any idea how the tides could lead our friend to prefer a
particular astronomy, so it is now your task to explain that.

Sarpi   One day in 1595, as I said, our friend came to see me in some
excitement. He began by saying that the periodic rising of huge
quantities of water, contrary to its natural inclination to remain
level, or to level itself quickly if disturbed, had greatly puzzled
him from the first time that he had seen this effect in our lagoons

D 425   at Venice. While crossing to Venice that day on a barge bringing
fresh water from the mainland, he had been struck by the large rise
of water near the prow when the barge scraped bottom in some
shallow waters and thereby lost headway in its forward motion. He
then saw that for some time afterward the water rocked back and
forth, rising and sinking alternately at prow and stern, but not
near the middle of the barge. The motion being leisurely and
regular, it next occurred to him that if the sea beneath had a
regular rocking motion, still more gradual, then its waters would

rise and fall against our shores, as indeed they are seen to do, being able to flow freely to and fro only in the central parts. In that case we would not have to look for distant forces to lift those huge and weighty masses of water, which forces would have to be very powerful. To start the rocking motion, all that would be necessary was for the seabasin, like the barge itself, to gain or lose speed from one time to another.

Sagredo   I remember that he did not like to assume forces for anything that could be explained by simple motions, saying that forces were too much like the occult qualities invoked by philosophers to explain any effect whose cause remained obscure.

Sarpi
D 462   Yes; that was his objection against the common opinion that the moon has power to attract and lift the waters of our seas. Much better, he said, was the explanation of a motion by some other motion seen to exist, as the lifting of water in the barge was explained by its own forward motion, effective when the barge was stopped or slowed.

Sagredo   But what stops or slows the basin of a sea?

Sarpi   That is, of course, what I asked him. "Nothing stops it," he answered, "but if the whole earth moves with the two Copernican motions, every solid part of the earth changes speed continuously throughout each day. Water, on the other hand, does not readily change speed, as shown by its flowing and rising against an obstacle, contrary to its nature as a heavy thing." He then outlined his explanation much as I wrote it down in my notebook, thus:

☆ GS 201   568.—The diameter of the earth subtends six minutes of arc, or a bit less, in the orbit of its annual motion, whence, its center moving less than sixty, that advances in the upper part thereof one-fifth in {an arc of} thirty [minutes], and in the other thirty it is retarded by that amount. The upper is night, and the lower, daytime. Hence every point on the [earth's] surface is now fast, now average, and now slow in the sphere [of fixed stars].[47]

As you can see, this was jotted hastily but is easy enough to follow. Sixty minutes of arc is a bit more than the center of the earth moves in one day in the Copernican annual orbit, since we have 365 days to travel 360 degrees, and sixty minutes is one degree. The upper part of the earth is that farthest from the sun, where it is night. Since both motions are in the same direction, they add together in the upper part, so that while a point moves 30 minutes in the annual orbit it advances six minutes, or one-fifth, in the diurnal motion. But in the lower part the annual motion is contrary to the daily rotation, which must therefore be subtracted from its speed.

Salviati   I see; he takes the absolute speed as that of the progress of a given point around the sphere of the fixed stars. During the twelve hours of "night," the point that started as westernmost becomes easternmost, as seen from the sun.

Sarpi   Of course; and, since in the Copernican astronomy the sun is a fixed star, we could say that the absolute speed of any point on earth is its speed as that would be measured by an absolutely fixed solar or stellar observer. Next:

☆ *GS* 201–2   569.—Any water carried in a basin, at the beginning of its transport, remains behind and rises at the rear, because the motion [of the basin] has not yet been thoroughly received; and when [transport is] stopped, the water continues to be moved by the received motion and rises at the front. The seas are waters in basins and by the annual motion of the earth make that same effect, being now swift and now slow and again average through the diurnal motion, which [effect] is seen in the moving of the basin diversely. And if seas are so large that they extend a quarter-sphere, so that part is in the swift and part in the average motion, there will be greater diversity, and still greater if they have a half-sphere, so that part is in the swift and part in the slow motion.

Salviati   Conversely, seas may have no perceptible disturbance if they are very small; or should I say, very narrow east and west, for some might be quite large but tideless, like the Caspian Sea.

Sarpi   Our friend did not overlook that, and he also took note of other reasons for differences of tides in different parts of the earth, and at different times, thus:

☆ *GS* 202   570.—Hence it is manifest that lakes and small seas do not produce this effect, being insensible basins. It is also manifest how variation of eccentricity, varying the ratios between annual and diurnal motion, equalizes the augmentations and decrements.

Salviati   Excuse me, but I do not understand that, let alone see it to be manifest. What does it mean?

Sarpi   The fault is doubtless mine. It is many years since I wrote this, and I see that I neglected to add "in our northern and southern hemispheres." The daily motion of rotation vanishes at the poles, and has different ratios to the annual motion at different latitudes or "eccentricities" in the sense of angles with the equatorial plane. The net daily speeds are equal at similar latitudes north and south. The note continues:

☆ *GS* 202   Also it is manifest how various positions of the shores may cause variation if their length be along or across the motion. Finally it is manifest how the motion of the [sea basin through our] seasons, carrying shores now to one site and now to another, makes an annual variation in augmentations and decrements.

   As you can see from the word "finally," that ended the outline of our friend's reflections when we first spoke of his proposed explanation. There is another paragraph, however, added a day or two later; for before our friend returned to Padua, I, having had time to reflect, asked him a further question. Perhaps you can guess what it was, and how he replied.

Salviati   It seems to me that there is one great difficulty with this fascinating speculation. Regular motions of our great seas are here explained in terms of continually changing speeds of every point on

D 262
D&O 265

the solid earth on the assumption that the earth has two circular motions, one around its axis, daily, and the other around the sun annually. That third motion in the Copernican system, related to the seasons, was also mentioned but need not detain us now.[48] Considering only the two chief motions, and recalling the reluctance of water to gain or lose speed immediately when its container is moved more rapidly or more slowly, we have a cause of continual disturbance in very large seas extending eastward and westward. But that disturbance would not be affected by anything remote from the earth, as is the moon.

Now, every writer on the tides says that they rise and fall with the motion of the moon, becoming greater or less with its changing phases. I do not know that those things are true everywhere, but they must be true of our Mediterranean Sea, which has been observed for centuries by those who have written on the tides, and all agree in saying this. I would therefore have asked our friend where the moon is hiding in his explanation of the tides. But I should like to know whether Sagredo, who shares your familiarity with actual tidal phenomena, thinks it reasonable that you would put that question rather than some other.

Sagredo   We who live at Venice, Salviati, have two sources of information about tides, one of which is direct observation of those conspicuous events here. That confirms in general what you have read about the influence of the moon on the waters of our sea. Our other source is conversation with sailors and sea captains who tell us of strange things seen in many ports remote from Italy. They do not always agree in the wonders they recount, but I have heard enough from them to make me question the common notion that some simple influence of the moon suffices to explain all tides in every sea. I would accordingly have asked a different question, though one that is closely related to yours and is perhaps more basic. Our friend's explanation, if I understand it correctly, seems to me to lead us to expect each day but one high tide and one low tide, whereas in fact we observe two tides—or rather, nearly two—each day.

Sarpi   It is Sagredo who has correctly divined my question. Let us defer
Salviati's puzzle until we have read our friend's reply to me, in
which (as you will see) he spoke only of our sea, and not of seas in
general as before:

☆ *GS* 202   571.—The motion of the sea will therefore be a motion composed
of two things: first, of the resting behind and rising of the water
[against one coast] when the motion passes from slow to swift; and—
from its natural returning to level—the second, of following the
motion when it [the water] passes from slow to swift and therefore
rises with its natural return to level.

Salviati   I see; in our friend's theory, the cycles of tides actually observed
depend not so much on the period of the general disturbance as on
the length of each particular sea from east to west. A very complicated interworking between disturbance, obstruction of flow by
coasts, and natural leveling of water, results in the tides we observe. In the Mediterranean, or at least here in the Venetian Gulf
at the end of the Adriatic, the cycle of rise and fall happens to be
made at about double the frequency expected from the general
disturbance alone. Is that the idea?

Sarpi   That is how he presented it to me some twenty years ago. You see
that your question regarding influence of the moon became of little
importance from a logical viewpoint. If we can accept a halving of
the general period of disturbance in virtue of particular conditions
of one sea, then a further difference of three-quarters of an hour
every day could very easily result from the same particular conditions. I remember that we talked of this, and our friend made a
very amusing remark. He said that lunar influence does not cause
D 445   the tides, but that tides cause lunatic theories—that is, they cause
simpleminded explanations to be accepted by people who have
never observed tides except in the Mediterranean and believe them
to be the same everywhere in the world.

Salviati   It is true that ours is the only sea that has been carefully observed
D 433   and written about for many centuries, and it could be that the

lunar-influence theory of tides originated from the common habit of generalizing from a single case. On the other hand, if the moon does indeed affect the tides, as is commonly believed, our friend himself may have been carried away by enthusiasm for a mistaken explanation of tides just because it was entirely his own invention.

Sarpi   If, for example, he had already been an enthusiast for the new astronomy, and had hit upon this tide theory in which he saw independent support for Copernicus, that could easily have influenced him unduly.

Sagredo   But you and I, Fra Paolo, who spoke with him often about that time, do not recall any previous enthusiasm on his part for the new astronomy, or even any great interest in astronomy of any kind, until the telescope began to provide him with new kinds of astronomical knowledge. His present confidence that the new astronomy will triumph appears to come from its having been confirmed independently by several things of which he did not dream when he first began to prefer it to the old system.

Sarpi   Returning now to what I said earlier, there is a way in which long study of natural motions of heavy bodies could lead to a belief in motion of the earth itself. Kepler provides an interesting example of the difficulty of the reverse order of interests. For Kepler, who believes passionately in the Copernican motions of the earth, resolutely refuses to believe that they could be the primary cause of tides. Instead, he ascribes to the moon a power to raise the waters of a sea as it passes over them.

Salviati   This is curious. A while ago I quite overlooked the double cycle of daily tides that Sagredo and you saw as the greatest difficulty with our friend's explanation. Now, recognizing that you were right, I take what I have learned from you and ask how Kepler, and other supporters of lunar influence, account for that daily double cycle, which you also tell me is quite precisely observed here, being lunar rather than solar in its periodicity?

Sarpi   Kepler, so far as I know, has never explained it at all, but some
D 420   astrologers and philosophers who ascribe to the moon a power of specific attraction for the waters of our seas grant to the moon also

the power to confer half of its power on the sign of the zodiac that happens to be directly opposite to it at any time.

Salviati   A benign and tractable philosophy indeed, but one that falls far short of science as we view it. Our friend's explanation requires very complicated motions arising from two independent causes, perhaps so complicated as to defy analysis in every set of particular conditions, whereas the rival explanation of lunar attraction, solves everything very simply and at one fell swoop. In choosing between the two, it seems to me that the first thing to do is to inquire further into actual phenomena. Are tide cycles the same everywhere, as we should expect from the simple theory, or do they greatly differ in different seas, as we should expect under our friend's explanation? If you will excuse the play on words, I am all at sea on this, having always lived inland, whereas you two have conversed with captains, sailors, and travelers.

Sarpi   For my part, I have reason to believe that tides are *not* the same in every sea and on every coast. In the Gulf of Aden, for example, I am told that for two weeks there is a daily cycle, and then for two weeks there is a double cycle, as here. On the coast of China, navigators say, only a daily cycle occurs in many large areas, though on the opposite coast of the Pacific Ocean, in America, the usual cycle is double, as on its Atlantic coast and in Europe. Still more remarkable is the absence of tides reported on the shores of islands in the middle of the Pacific Ocean. There are a great many islands of considerable size that are said to be tideless, or very nearly so.

D 433

Sagredo   I have noticed during my many voyages in the Mediterranean that islands far from the mainland, such as Malta, have no tides. That seems to support our friend's analogy of tides to the rocking of water in vessels, in which it rises and falls at the ends but not in the central parts where the water is free to flow unobstructed.

Salviati   This seems to impose a formidable objection against the lunar-attraction theory. If the moon does raise the water beneath it, it should do so on the coasts of islands just as on mainland coasts to the north or south of them.

Sagredo   Not only does it not, but in talking with captains during my voyages I learned something curious about the tides as related to the moon. As you say, if a lunar bulge of water travels westward under the moon, it should create tides at the same time on the northern and southern coasts of the same sea, but that is not seen. In places where tides are great, it is important for captains to be able to calculate when high tide will next occur at a port they are approaching to discharge cargo. For that purpose they carry tables of what is called the "establishment" of each port. It is true that high tide occurs at any port always when the moon is in a certain position with respect to the meridian of that port. That position is called "the establishment of the port," and it may differ by an hour, or two, or even three from one port to another. Now, the curious thing is that those differences are not observed in an orderly fashion along a coast running east and west, as for instance along the English Channel. I have been told that at Dover and Dieppe the port establishments differ by an hour and a half, as if the moon attracted the water from overhead at one port but from a position fifteen hundred miles distant east or west at the other port. Yet those ports hardly differ in longitude.

Salviati   Dear me, this matter of tides becomes so complex, when we even begin to take into account what navigators know about them, that many different causes may have to be taken into account in truly explaining tides. That tends to favor our friend's approach, in which he at least recognizes two separate causes, one for continual disturbance of the seas and another regulating their observed periods. And now I think I begin to see how a certain difficulty may account for our friend's reticence about this important matter since he came to Florence, and how that difficulty may account for his long delay in publishing on the system of the world, even apart from his necessary expenditure of time in purely astronomical investigations that I have explained to you.

Sagredo   Do not stop at this hint; to what "certain difficulty" do you allude?

Sarpi   First let me guess, and then Salviati may say whether I am correct. I believe you refer once again, Salviati, to the complete absence of the moon from our friend's explanation of tides.

Salviati	You are exactly right. Let me counter by asking on what grounds you guessed this, for I am greatly enjoying our little contests of this kind as we go along.
Sarpi	We have seen that serious objections exist to any simple ascription of attractive power to the moon, directed specifically to draw up the waters of the sea beneath, as an acceptable explanation of the observed tides. Nevertheless we have seen at the same time that to deny any role whatever to the moon is imprudent. It would be permissible to say that the length of our Mediterranean only happens accidentally to occasion reciprocations of its tides that coincide with returns (or half-returns) of the moon, if we knew no more than what can be observed here at Venice. Doubtless that seemed sufficient to our friend when he first hit on his idea, for having (like you, Salviati) been an inlander all his life, he probably knew no more about tides than what Strabo wrote. But it would be quite another thing for him, having learned that the moon is somehow, if confusedly, linked with the "establishments" of ports, to reject lunar influence of *some* kind as active in tidal phenomena. I believe you meant that the difficulty of finding *any* role for the moon is still troubling our friend.
Salviati	Precisely. He would, I think, be unwilling to abandon his original explanation entirely, especially now that he has other and very different reasons for believing in the doubly circular motion of the earth. But he might prefer not to disclose his system of the world without including so principal a phenomenon as the ocean tides. He would see this remaining defect not as destroying his explanation, but as needing to be removed by further study.
Sarpi	That is what just occurred to me, and if we are correct, it supplies a very clear and powerful explanation of the delay that so puzzled me when we began our conversations. In a way this strikes me as even more convincing than the fact that our friend has been busily occupied with other astronomical inquiries, very time-consuming by reason of many observations and innumerable calculations. My reasoning is this. If he were to omit from his promised book some astronomical information obtainable by observation and calcula-

tion, that would be a matter of no great consequence, since other astronomers would in time unfailingly supply the omission by pursuing the methods followed for centuries in their discipline, employing the instruments with which he has newly enriched it. But if he were to omit so significant a discovery as scientific explanation of tides, linking astronomical phenomena with terrestrial physics, centuries more might elapse before any such linkage would even be suspected.

Sagredo   I see why you say this. The notion of such a linkage, contrary to the basic postulate of Aristotle that the science of motion must separate celestial from earthly phenomena, came to our friend accidentally as the result of his observing in a barge some phenomena analogous to events in the water beneath the barge. That accident could not have happened at Florence, any more than the same observation made at Venice by innumerable bargemen and passengers ever inspired in them a similar reflection. It might indeed be centuries before any other person would be led to it.

Sarpi   You must add "in the same way," Sagredo. What has kept astronomers from linking their science with physics is the authority of Aristotle and the ingenious compromise proposed by Geminus that allowed them to continue undisturbed in their investigations only at the price of surrendering physics entirely to philosophers. But Kepler, despite the admonitions of his teacher Michael Maestlin, dared to defy tradition and explore the possibility of a celestial physics in his *New Astronomy* four years ago. It is there that we find Kepler's proposed explanation of tides, so I believe that he, and other daring spirits like him, would in time have found the truth about tides even without such a happy accident as inspired our friend.

Sagredo   Yet Kepler denied that motions of the earth could explain tides and
x, 72   proposed a theory that we have seen to be inadequate to account for the observed cycles, or for the absence of tides on mid-sea islands.

Sarpi   You are unfair to Kepler, because we have had occasion only to consider one aspect of his theory, which may be inadequate and indeed almost laughable, presented alone. Now, what Kepler proposed was that if the moon were not held in its orbit, it would be drawn to the earth and the earth would be drawn to it by their natural inclinations to motion, and they would meet at a place determined by their respective sizes, the larger earth moving more slowly than the smaller moon. Being held apart, they cannot move toward one another, though the fluid seas betray a natural inclination of *all* bodies to do so.

Sagredo   I grant that on such an assumption the tides would become part of a celestial physics, and since Kepler is an avowed Copernican, they would in his view be at least indirectly linked with motions of the earth. Yet Kepler's explanation of tides is very defective, even apart from that double cycle that he neglected. I might even say that it is self-contradictory.

Sarpi   In what way?

Sagredo   Anyone who held the earth to be immovably fixed at the center of the universe could reasonably allow only the seas to betray a supposed attraction by the moon. But Kepler, who allows the earth to move, would have to explain why the whole earth is not moved by the moon just as much as are the seas, in which case the waters would not be lifted at all with respect to their basins.

Salviati   Let us just say that neither Kepler's nor our friend's explanation is adequate, from what we know of their rival tide theories at the present time. Perhaps one or the other can be adequately amended and supplemented; perhaps not. It may even be that motions of the earth and a universal attraction of bodies to one another are both needed to explain the tides. This whole discussion has been interesting, but we should move on to other matters, unless anything else relating to the tides occurs to you that we have not considered, and that might be relevant.

Sarpi   There is one thing we have neglected, though I do not see its bearing on our friend's perplexing problem. This is that in addition to the moon's role in the establishment of a port, it has an indubitable influence on the seas that cannot be charged off to particular local conditions. This is that at new moon and full moon, or a day or so

later, the tides at any port rise and fall more, and at first and last
quarters they rise and fall less, than at other times of the month.

Salviati   A remark about that by our friend, that I heard only casually and
paid no further attention to, came to my mind when you first
mentioned his having had a theory of the tides linking them with
motions of the earth. One day, when we were talking about some
curiosities of light and illumination, he said: "If only I could
understand why tides are greater both when the moon is most
brightly illuminated and when it is quite dark. It does seem that
if the moon can affect tides, as everyone believes, then its influence
should wax and wane with some other observable phenomenon of
the moon herself, rather than as it actually does." As I said, we
were talking about light, not about tides or motion of the earth,
so I did not ask him why he wondered about this. Perhaps he has
been seeking to connect tides with phases of the moon through
some kind of a secondary cause, of which, as we said, there are
probably many more that must be taken into account in any complete understanding of tides.[49] But now let us hear from Sagredo
about the connection of planetary speeds with natural motions of
heavy bodies, which would be no less a linkage of astronomy with
physics.

Sagredo   I shall tell you what little I can. But first, since I have the *Sunspot
Letters* at hand, I want to read another passage that, though written
about sunspots, reveals an attitude of our friend's that would
equally apply to these perplexing problems of the tides:

☆ *D&O 90*   The difficulty of this matter, combined with my inability to make
many continued observations, has kept (and still keeps) my judgment
in suspense. And indeed I must be more cautious and circumspect
than most other people in pronouncing on anything new. As Your
Excellency [Mark Welser] well knows, certain recent discoveries that
depart from common and popular opinions have been noisily denied
and impugned, obliging me to hide in silence every new idea of mine
until I have more than proven it. . . . I might add that I am quite
content to be last and come forth with a correct idea, rather than get

ahead of others and be compelled later to retract what might indeed have been said sooner, but with less consideration.

Sarpi   I only hope that this prudent rule will not induce our friend to wait too long for enlightenment, and thus to carry to his grave these possible connections between astronomy and physics. Do tell us now about that other link, Sagredo, that Salviati wishes to hear before we conclude our discussions by pooling our ideas about the nature of our friend's long-delayed book.

Sagredo   By an amusing coincidence, his speculation about planetary speeds began from our friend's dissatisfaction with Kepler's first attempt to envision a celestial physics. I believe it was in 1597 that he received from Kepler a copy of his first book, in which the speeds of planets were discussed in relation to their distances from the sun. Kepler supposed that some kind of force extended outward from that center, capable of driving the planets around in their orbits. This force was assumed by Kepler to weaken with distance, so that more distant planets moved around more slowly. Five years later our friend, continually investigating motion and mechanics, conceived—as he later wrote in his book on bodies in water—that speed and weight are so related in heavy bodies that the *moment* of a

CES 31   rapidly moving small body can equal that of a slowly moving large body. In balances and levers it is seen that *moment* is increased with the distance at which it acts in driving a weight around the center.

GAW 63   Dissatisfied with Kepler's reasoning about a solar force, our friend took Kepler's measurements of planetary speeds and distances from the sun, inquiring whether his own principle of *moments* might be applied.

Salviati   But in his view planets are neither heavy nor light, since they
GAW 294   move naturally neither toward nor away from any center.

Sagredo   That is true, but he asked himself whether speed might not entirely *replace* weight, since increased speed can exactly compensate
GAW 64–65 for deficiency in weight, as I said. Out of this investigation came a remarkable relation between the ratios of distances from the outermost planet, Saturn, to each other planet, and the ratios of

	each planetary speed to that of Saturn. The former ratios were, so to speak, the square roots of the latter, or, as mathematicians say, as their subduplicates.
Salviati	Most remarkable, but I do not see how that has anything to do with the descent of heavy bodies, or even with distances from the sun.
Sagredo	Our friend did not attempt to relate them at the time, because he had not yet discovered that law governing speeds of heavy bodies in natural descent of which Fra Paolo spoke. Perhaps you will say something more about that now, Fra Paolo, after which it will be easier for me to finish my story.
Sarpi	Certainly. Among the many problems of natural motions we used to talk about, one of the most perplexing was that in order to get from rest to any speed, a heavy body dropped would either have to move instantaneously, so to speak, and start falling with some speed that was somehow natural to it, or else it would have to pass through every possible speed in its fall, which speeds would be infinitely many. Neither alternative seemed satisfactory. After we

GAW 86–90 had long debated this problem, our friend thought of a way to
measure times and distances in natural descent. To slow down the
motion, he used a very gently sloping plane, grooved to guide a
bronze ball rolling down it. By marking its places at the end of
equal intervals of time, he found that the distances from rest had
the same ratio as the squares of the times elapsed. This rule for
motions of heavy bodies descending naturally toward the center of
the earth was found to be true for planes of different slopes, and, as
nearly as our friend could tell, for straight fall also, though that is

GAW 102–3 too fast for precise measurement. The year was 1604, for I remember that he wrote out a mathematical proof of it for me about the
time that the new star appeared.

| | |
| *Sagredo* | And four years later, shortly before I left Venice to serve our Most Serene Republic in Syria, our friend told me that this same law seemed to be connected with the speeds of planets.[50] He had not noticed this before, because his first assumption about speeds, used in the proof just mentioned by Fra Paolo, had been mistaken. |

D 21, 29

When that was finally corrected, the rule of speeds in fall was applied in a certain way to planetary speeds. Out of curiosity, our friend tried the hypothesis that all the planets had acquired their speeds by descending from one remote place toward the sun, and he concluded that if each planet had been turned by divine will to circular motion at its present distance from the sun, it would have acquired very nearly the speed in orbit now observed.

Salviati   What a marvelous conception! Certainly that would induce any reasonable person to regard the sun as much more probably the center of the planetary motions than the earth. In any event such a discovery, even if only approximately confirmed by calculations and measurements, will surely find a distinguished place in our friend's promised book about the system of the world. It calls to mind Plato's story in the *Timaeus*, according to which after primordial wandering in chaos, the world bodies were brought into order by God's will. But I do not believe that it could be confirmed more than very roughly by the best information taken from modern astronomers.

Sarpi   Let me add my bewildered delight at what you say, Salviati, and then ask why you append your final caution. Sagredo, if nothing else had been accomplished in our discussions, which is very far indeed from being the case, this remarkable speculation of our friend's would alone crown them with delight for me. I only wonder that he did not confide it to me, as he did to you.

Sagredo   I do not think that there is any great puzzle in that, Fra Paolo. It is one thing for a scientist to hit upon some mathematical or physical proposition evidenced by observations and measurements, but quite another for him to venture into the domain reserved for theologians, which includes the mysteries of the Creation. I myself am surprised to hear you welcome this discovery without hesitation. Might it not raise some profound theological issue, for which reason our friend might have feared your displeasure?

Sarpi   One, yes, now that I stop to think of it. The Bible does appear to assert creation of the earth before that of the sun, though that account of the events is so brief as to be perhaps only metaphorical.

Probably you are right in supposing that our friend withheld this discovery from me lest it disquiet me as rash or even erroneous in our Catholic faith.

Salviati   But as Copernicus wrote in his dedicatory epistle to Pope Paul III, "mathematics is written for mathematicians," and this speculation was purely mathematical. Nor did our friend hesitate to impart to you his explanation of the tides, which was based on motions of the earth. That is no less offensive to many theologians.

Sarpi   The cases seem to me quite different in two ways. First, explanation of great motions of our seas is clearly a proper problem to be solved scientifically if possible, whereas the mystery of Creation is not. It would be of great use to mankind if we could predict the motions of our seas, while no such utility justifies speculation about the Creation. And second, some passages in the Bible seem to speak of motion of the earth, cited by the Spanish theologian Diego de Zuñiga in his commentary on *Job* to support the Copernican astronomy, opposed by other theologians. But no contrary opinions in dispute among theologians are based on biblical accounts of the Creation.

Well, since we cannot solve the problem of our friend's reticence toward me in this matter, let me pass on to the other question. Why did you say, Salviati, that only rough agreement can be expected between his speculation about planetary speeds and distances, and the best measurements to be had from astronomers?

Salviati   Here, Fra Paolo, we shall be entering into a principal matter about our friend's system of the world as I understand it at present, so let us spend a little time on your question before we attempt to sum up our long discussions. To begin with, we all know the scheme of the heavens depicted by Aristotle, in which each wandering body and all the fixed stars circle the earth as the unique center of celestial motions. That arrangement was worked out by Eudoxus at Plato's suggestion, so it is not in dispute among philosophers. On the other hand we know that the best measurements of astronomers have conflicted with it almost from the beginning. To describe the motions actually observed, astronomers had to introduce eccentric circles and epicycles for the sun, moon, and planets, by which their distances from the earth and their speeds in orbit were rendered not constant, but variable.

Sarpi   I have read and heard many discussions of this among philosophers. They seem to be chiefly of the opinion that those devices introduced by astronomers (as mere mathematicians) are simply fictions not to be taken seriously, the true celestial motions being uniform and circular around the earth, as both Plato and Aristotle supposed.

Salviati   That was the position taken by La Galla, and I remember that our
iii, 338   friend jotted a note saying that truly concentric orbits for the planets were simply false and impossible. Every astronomer knows
D 453, 455   that, and Copernicus no less than Ptolemy made use of eccentric circles and epicycles. By assuming the sun rather than the earth to be the center—or rather, nearest of all bodies to the center—of planetary motions, Copernicus reduced the epicyclic alterations of distances from the center in size, but the impossibility of precisely concentric circular orbits remained.

Sagredo   Now I begin to see, Fra Paolo, why Salviati said that only rough agreement could exist between our friend's fascinating speculation and the best astronomical measurements. By his calculations from that law of speeds in fall, a unique distance from the sun and a unique speed would result for each planet. Astronomical measurements, on the other hand, require changing distances and changing speeds of planets, as long as epicycles are required to describe the observed motions.

·   *Salviati*   What Sagredo says is true, and the situation is even worse than that. Even without epicycles it has been found impossible to reconcile uniform circular motions of planets with those actually observed. I refer to the recent book by Kepler in which epicycles are abandoned, but only at the price of adopting elliptical orbits in which planetary speeds are not uniform, but continuously changing. Now, the astronomical measurements used by Kepler were the very best, made by the extremely accurate Tycho Brahe. No hope therefore remains that precisely uniform and circular motions, as required by philosophers on metaphysical principles, will ever be confirmed by actual measurements carried out by competent astronomers.

Sarpi At last I see clearly why our friend, in announcing his fascinating discovery to Sagredo, told him that the calculations answered only "very nearly," a phrase that would allow considerable leeway. Still, knowing our friend, we are safe in assuming that he had found sufficient agreement to persuade him that it was not merely coincidental, so that some relation probably exists between the law that governs speeds in fall of heavy bodies near the earth and the speeds of planets in their courses. That would confirm your earlier remark, Salviati, that the promised book will contain not only astronomical but also physical reasoning, and perhaps even more of the latter than the former.

Salviati Yes, I think that with the information we have just had from Sagredo there can be no doubt of that. And as a consequence our friend's delay is not at all surprising, for he will have to be very cautious in mingling physics with astronomy—against the precepts of philosophers and the compromise of Geminus, long gladly accepted by astronomers.

Sagredo Clearly he will have to answer all objections that have been raised against the possibility of physical motions of the earth, which are

D 126–27 many and seemingly cogent, especially some that were added by Tycho to the ancient ones of Aristotle and Ptolemy. It seems to me that what Salviati told us yesterday sufficiently explained delay of the promised book in terms of the time our friend has had to spend on astronomical problems, such as determination of the satellite periods and investigation of sunspots, even without this new reason for proceeding very cautiously in writing on the system of the world. If Fra Paolo agrees, we might well now go beyond our original goal of understanding reasons for that delay and try to guess at the probable content of the book when it does appear, considering problems that still detain our friend and ways in which he may attempt to solve them.

Sarpi I do agree, Sagredo, for many reasons. The first is that in a way that would not alter our original goal, but would give it new meaning and purpose. When I first asked Salviati what was holding our friend back, I wondered not just about his other activities, of which we have since been told, but also whether perplexities and doubts about the system of the world might have caused him to abandon the project. Shall we turn to such matters as Sagredo proposes, Salviati?

Salviati   Let us by all means do so while we are all three together, each knowing different facets of his inquiries and his habits of mind. Only this morning have I become convinced of what I began to suspect early in our talks—that profound problems of physics, of which our friend has hardly hinted since his move to Florence, must be resolved before he will venture to publish on the system of the world. We may profitably pool our ideas about those with relation to the larger problem of the arrangement and motions of heavenly bodies.

Sagredo   A good place to begin, then, it seems to me, is with those motions themselves, of which our friend said as little here at Padua as you tell us he has said about his new science of motion recently at Florence. We know that as to the constitution and arrangement of the heavens, he adheres, with certain qualifications, to the Copernican view. As you remarked a while ago, Salviati, Copernicus was no less obliged to introduce epicycles and eccentrics than was Ptolemy, though very differently. Fra Paolo said that philosophers generally dismiss such motions as mere fictions. On so central a question our friend will have to take some position in writing on the system of the world. I do not quite understand his recent statements about this in the *Sunspot Letters*.

Salviati   That is strange, for I thought them quite clear. Perhaps you should first refresh our memories by reading them, for you have the book right there.

Sagredo   Let me look for the passage . . . yes, here it is. Commenting on Father Christopher Scheiner's opinion as to the location of sunspots, our friend wrote:

☆
D&O 96–97   In this he continues to adhere to eccentrics, deferents, equants, epicycles, and the like as if they were real, actual, and distinct things. Those are postulated by pure astronomers in order to facilitate their calculations, but they are not to be retained as real by astronomer-philosophers who, going beyond the task of somehow saving the appearances, seek to investigate the true constitution of the universe as a principal and most admirable question. For such a constitution exists and in but one way, true, real, and impossible to be otherwise; and the greatness and nobility of this arrangement entitle it to be placed ahead of every other question knowable by speculative minds.

Thus far it appeared to me that our friend agreed with the philosophers who regard eccentrics and epicycles as mere fictions, since he said that astronomers who wish to philosophize about grander problems should lay them aside, and surely he would not advocate the laying aside of facts. But, to my perplexity, he went on to say:

☆ *D&O 97*   I do not deny the existence of circular motions about the earth and upon other centers than the earth's, or even other circular motions completely separated from the earth that do not go around it, and remain within their own circles. The approaches and retreats of Mars, Jupiter, and Saturn assure me of the former, while Venus, Mercury, and the four Medicean planets render me certain of the latter. Hence I am quite certain that there exist circular motions that describe eccentric and epicyclic circles.

This appeared to contradict my former conclusion, since our friend now says that he is sure of epicyclic motions as facts. But he then . . .

Salviati   Pardon me for interrupting, Sagredo, before you have finished. I wish to point out, for discussion later, that what you have just read shows that our friend does not accept the elliptical orbits

expounded by Kepler, from which epicycles disappear as super-
fluous. Rather, he speaks of circular motions in the sense of
Copernicus, or of Ptolemy, though not in that of Aristotle, or of
Plato. Please continue, and excuse my interruption.

Sagredo   Unless we freely interrupt one another as we enter this complicated
business, much may be forgotten or lost to view that might help
to clear it up. Our friend then went on:

☆ D&O 97   But that Nature, in order to provide these, really makes use of that
farrago of spheres and orbs depicted by the astronomers is, I think,
hardly to be believed, as something accommodated to the facilitation
of astronomical calculations. My opinion lies midway between that of
those astronomers who assume not only epicyclic movements of the
[wandering] stars but also orbs and eccentric spheres which conduct
them, and the opinion of those philosophers who deny both the orbs
and also movements around any center other than that of the earth.

Before we discuss what opinion can lie midway between acceptance as facts and rejection as fictions, I should mention that the
printed text says "eccentric" where I just read "epicyclic," suspect-
ing a slip of the pen or a printer's error.

Sarpi   Certainly you are not to be blamed, Sagredo, for remembering
these statements as embodying a contradiction and ending with a
paradox. It is to the apparent paradox that we should look for
resolution of the seeming contradiction. Some astronomers do
indeed assume, without necessity, the reality of orbs and eccentric
spheres, when all that is required from astronomers is description
of the movements observed. Our friend rejects the unnecessary
assumption of such astronomers, saying that it should be left
behind in philosophizing about the true, real, unique, and necessary arrangement of the universe. Thus far there is no contradiction
between the first part of the statement and the last; or do you not
agree?

Sagredo   Thus far I agree with you; what next?

Sarpi   Next, when our friend says he is sure that epicyclic motions exist in the heavens, let us consider just the motions of the four Medicean stars and neglect the planets for the moment. Jupiter's satellites certainly move in epicycles, whether seen from the earth or from the sun. Seen from the earth, the center of their motions is carried on a very eccentric circle, but seen from the sun their deferent circle is very little eccentric, perhaps negligibly so.

Sagredo   You pause as if I were to reply; please continue, for there is no difficulty in this.

Sarpi   Then you see now why our friend's opinion is not that of the philosophers (and many astronomers) who deny celestial motions around any center other than the earth's. He accepts such motions, although he abandons the orbs. His position is midway, accepting something from astronomers and something from philosophers but agreeing with neither wholly.

Sagredo   I see, and now the rest is clear. Each adversary party has brought into philosophizing about the true constitution of the universe something useful and something else superfluous that should have been left behind. The orbs and spheres, postulated by some astronomers to facilitate their calculations, need not and should not be deemed physically real and in the heavens; those are mere tools used necessarily in building a structure, not part of the completed structure. But the movements so calculated cannot properly be ignored by philosophers, as is done when they bring in from philosophy the superfluous principle that all heavenly motions circle the earth, an assumption they should have left behind. Our friend stands with his astronomer-philosophers, who know what to bring to the larger task and what to leave out.

Salviati   Or rather, he stands among them. When you said "stands with them" Sagredo, you spoke as if all astronomer-philosophers are in agreement, which is not the case. Disputes and disagreements exist among them just as they do among pure astronomers or among pure philosophers. Kepler is certainly an astronomer-philosopher, and of the highest order, as our friend will readily acknowledge; yet he and Kepler disagree about astronomical as well as physical conclusions, such as shapes of orbits and causes of tides.

Sagredo D&O 96, 103	For that matter our friend seems to accept even Scheiner as an astronomer-philosopher, "deeming him a person of high intelligence and a lover of truth" in his first letter replying to Mark Welser. But judging from the errors he exposed in Scheiner's opinions, I think our friend might reverse the terms and call Scheiner a philosopher-astronomer.
Salviati	A nice distinction; but tell me why you think our friend recognizes Kepler, and not Scheiner, as a fully kindred spirit?
Sagredo	With one grave misgiving, I should answer at once that our friend and Kepler have in common a deep respect for precise measurements as the key to philosophizing correctly. Scheiner lacks this, for he did not change his opinion when our friend adduced careful measurements of sunspot motions to place them on the very surface of the sun itself.
Salviati	I agree with you so wholeheartedly about our friend and Kepler that I should like to hear your misgiving in stating it.
Sagredo	It stems from a remark of yours a while ago, that our friend does not accept Kepler's elliptical orbits. Now, Kepler discovered those by years of labor in calculating from the most exact measurements, as he did not fail to emphasize in his *New Astronomy*. It seems to me that if deep respect for precise measurements were all that distinguished astronomer-philosophers from philosopher-astronomers, our friend would welcome Kepler's elliptical orbits as freeing astronomy from the nuisance of implausible epicycles, to say nothing of the somewhat lesser nuisance of unexplained eccentric motions.
Sarpi	What! Did Kepler vanquish both those ancient blemishes from pure astronomy? I am no astronomer, so enlighten me.
Sagredo	Elliptical orbits for all planets, when we count the earth also as a planet, produce, with continually varying speeds in orbit, all the appearances formerly saved by epicycles and eccentrics. Moreover, by placing the sun not at the center of each ellipse but at one of the two points called *foci*, mathematically precisely determined in ellipses, Kepler also removed that ancient fiction of the equant which so exasperated Copernicus.[51] No fictitious empty point, but

a place occupied by the sun itself, serves for Kepler's ingenious calculations.

Sarpi   Now I am all yours, Sagredo, and Kepler's too. It is up to Salviati to explain how our friend could resist such blandishments, based on careful measurements.

Salviati   The answer brings us to the heart of things, if we wish to conjecture not just on the reasons for our friend's delay in setting forth his system of the world, but on the nature of his book when it appears. I do not know whether to begin with the attendant problems of physics, which have been already mentioned several times, or to start anew with the fundamental problems of actual measurement, which in a way lie even closer to the heart of the matter. Both have to be discussed; which shall we take up first?

Sagredo   Since we are now talking about Kepler, whose *New Astronomy* pioneered a celestial physics and motions "causally considered," let us begin with that, if Fra Paolo agrees.

Sarpi   Being less informed and having more to learn, I shall be guided by you, Sagredo.

Salviati   Very well. The great virtue of Aristotle's assumption about celestial motions was that by making those uniform and circular, he avoided the necessity of introducing any cause for them except nature itself. Any change in speed or in direction of motion by a planet would have to be explained, and Kepler, of course, realized this when he presented his evidence for elliptical orbits at changing speeds. Long before that he had already postulated a solar force to drive the planets around, which, as you remember from what Sagredo told us, did not appeal to our friend. To this, Kepler has recently added certain magnetic influences that alternately increase and decrease distances of planets from the sun, strengthening or weakening the force driving them around. Such, very roughly, is the celestial physics of Kepler.

Sarpi   Unless those forces were deduced numerically from measurements—and I do not see how they could be—I can easily understand why our friend would be reluctant to accept this. It would

	strike him, I believe, as very much like the orbs and spheres of ancient cosmologers, carried along as excess baggage into most philosophizing about the system of the world.
Salviati	Exactly. Kepler's celestial forces could not be deduced from measurements but could only be made to fit elliptical orbits ad hoc, as epicycles and eccentrics were long ago fitted ad hoc to circular orbits. Hence our friend could either accept elliptical orbits and devise a better physics to fit them, or else ignore them despite the fact that ellipses *were* deduced from exact astronomical measurements.
Sagredo	It is not hard to understand why our friend could not frame his physics to fit elliptical orbits. I have studied enough mathematics to know that calculations for ellipses are very complex, so I suppose that physics involving them could not be simple in its assumptions.[52]
Salviati	I should perhaps have mentioned earlier a second virtue of Aristotle's assumption for astronomers. Calculations for circles, and uniform motions in them, are quite easy. By combinations of such motions, especially with eccentric circles allowed, astronomers could deal systematically with all the observed motions in the heavens. Now, Sagredo, you imply that physics should adhere to simple assumptions, so what would you expect to find in our friend's promised book in the way of a celestial physics?
Sagredo	Certainly nothing relating to elliptical orbits. Perhaps nothing at all, but just terrestrial physics capable of overcoming objections to the earth's motion. Yet it seems to me that our friend would write something relating to the physics of planetary motions, in view of his speculation about their speeds.
Salviati	We had a hint in what you first read from the *Sunspot Letters* about the way in which most objections to the earth's motion could be met. There, you recall, it was said that bodies placed in motion to which they have no natural repugnance would forever remain in that motion if unimpeded externally, and thus a ship would continue westward, for example, if not obstructed. Most objections to the earth's motion assume with Aristotle that its rotation would be

revealed by observed deviation of falling bodies from natural straight motion, which would not be perceptible under our friend's rule. On the other hand I do not see how his rule could explain the circling of planets, or any other unsupported bodies, even if they acquired their speeds by first falling toward the sun as heavy bodies fall toward the earth.

Sarpi

M&M 74

In our friend's Pisan treatise on motion, mentioned earlier, he said that speed is increased by approach toward the center and decreased by greater distance from that. It would follow that constant speed could be maintained only by a body remaining at the same distance from the center. If by divine fiat a planet remained at the same speed acquired during fall toward the sun, it would move forever circularly around the sun by this other rule. That would agree approximately with the Copernican astronomy.

Sagredo

I see what you mean, and now I recall that our friend said he was sure that epicyclic motions exist in the heavens, as exemplified by the Medicean stars. All that would be needed for our friend's agreement with Copernicus, then, would be acceptance of the Copernican epicycles for planets. Those are not very large—nothing like the Ptolemaic epicycles—and though they would remain unexplained physically, the motions would be nearly borne out by measurements, unlike Kepler's solar force and magnetic attractions and repulsions.

Salviati

These conjectures suggest a possible approach to celestial physics in our friend's book on the system of the world, if that went beyond the terrestrial physics necessary to remove old puzzles. But I agree with Sagredo that our friend would not urge it, or do more than hint at it, perhaps only in presenting his discovery about planetary speeds.

Sarpi

D 32

One way occurs to me in which he could do this and still leave the door open to some other celestial physics, if time should vindicate Kepler's elliptical orbits. This would be to present celestial physics *ex suppositione*, as logicians call it. *If* any natural motion is uniform and perpetual, *then* it must be circular. It would suffice for our friend to say that, without asserting that any natural motion *is*

perpetually uniform—of the planets or of anything else. Then if
Kepler is right, and the planets continually change speed in their
courses, those courses might indeed be elliptical without contradicting anything our friend would have written. That is the virtue
of the method of Ptolemy, of which we spoke at the beginning
of today's discussion. Suppositions made to fit appearances may
always be altered with new knowledge.

Salviati   That would certainly be the most prudent way to handle such a
difficult and uncertain subject as is physics of the heavens at this
stage of our knowledge. It will be a long time, at least many
decades, before astronomical tables based on Kepler's orbits are
even compiled, let alone put to the test of prediction and observation of planetary positions. Meanwhile we are at the very beginning of an age of unparalleled precision of astronomical observation, thanks to the telescope. Whether elliptical orbits are in fact
enough more reliable for prediction than Copernican epicycles to
justify the enormously greater labor of calculating them must
remain to be seen. Now, that in turn brings up the related question
of actual measurements, which I said must be dealt with in order
to understand our friend's view of correct philosophizing about
nature. Since I believe we have said enough to suggest the form in
which physics may enter into his book on the system of the world,
let us now take up the problems of measurement.

Sarpi   I have been awaiting that with interest. Until I met our friend, I
was not even aware that fundamental problems exist about measurement, which seemed to me quite simple and merely tedious to
carry out with any precision one might wish. When our friend
began making careful measurements of motion, however, I became
aware of many kinds of real difficulty, even in being sure just what
was being measured. Hence I should like to hear why you brought
the subject up earlier and what you have to say about it.

Salviati   Your request recalls to me the start of our conversations, when I
tried to distract you from your principal question by asking you
about our friend's work here on motion. I brought up the topic of

measurement because philosophers often frame their principles in such a way that perfect agreement with Nature seems necessary. But perfect agreement is seldom or never found in actual measurements when those are made as precisely as possible. Indeed, when different people measure the same thing, and even when one person makes several measurements of the same thing, perfect agreement to the smallest observable unit of measurement is seldom found. Now, when philosophers reject evidence from actual measurement because it conflicts with some principle of theirs, either the principle or the measurements may be defective, so unless we understand the nature and limitations of measurement in the concrete, we may go astray in reaching abstract conclusions. How this bears on the system of the world is what I intended to discuss. But your remark that one may not even know what he is measuring puzzles me. Can you give us an example of what you had in mind?

Sarpi   This time I shall yield and attempt to satisfy you, as I would not do on that first day, when I wanted to hear news of our friend and you wanted me to tell you of his work on motion here years ago. As I said before, he was forced to resort to careful measurement because he could not decide by logic just how a body begins motion in natural descent. What he set out to measure with care were successive downward speeds in equal intervals of time from rest, but what he discovered was a ratio between distances and the squares of times. When he first tried to derive that ratio mathe-

GAW
98–102   matically by reasoning from a different measure of speeds, he fell into error, as Sagredo told us. The reason was that this different measure only seemed to be of speeds and was really of something else that depended on the squares of speeds. That is why I said that one may not even know what one is measuring, though the measurements are made with care.

Sagredo   He did not tell me all this, but only that some error had long kept him from discovery of that remarkable rule about planetary speeds. Please explain how our friend was misled about what he was measuring.

Sarpi   He was misled by sound reasoning, something that we discussed
            earlier as happening frequently in science. When a heavy weight is
            dropped on a resisting object, its effect is doubled if the height
            from which it is dropped is doubled. Our friend reasoned that
            since the weight remains the same, and only the speed changes,
            the falling weight must double its speed in a doubled height of
            fall. But in fact to double its *speed* it must fall four times as far as
            at first. In this case what our friend thought was a measure of
            speed at the end of fall was really a measure of impact, which is *not*
            simply proportional to speed. Thus an error flawed his first proof,
            in which the word "speed" had been used to designate two
            different things.

Salviati Thank you, Fra Paolo, for this unexpected light on a basic problem
            of actual measurement. The problem with which I was concerned
            is very different indeed, but no less capable of leading people far
            astray who suppose, as you say you did before you met our friend,
            that it is never difficult, but only tedious, to measure things to any
            desired degree of accuracy—even when knowing exactly what is to
            be measured. Philosophers who are not astronomer-philosophers
            commonly suppose the same. Thus, for example, they attempt to
            refute Copernicus by simply asserting that if the earth circled the
            sun, we who live on it would not always see exactly one-half the
            entire sky, as they believe that we do. Our friend used to tire
            himself out by patiently explaining the fallacy to people whose
            knowledge of geometry was minimal and who for other reasons had
            already closed their minds on the issue.

Sarpi   Now I remember that he once showed me a very long letter he had
            written to his former colleague Jacopo Mazzoni on this very ar-
ii, 198–202 gument, which Professor Mazzoni had uncritically adopted in a
            book, I believe in 1597. I take it that our friend now has a dif-
            ferent and less tiresome way of replying to philosophers who are
            not astronomers?

Salviati Yes indeed. He now makes use of the fact, known to all astrono-
            mers, that actual measurement can be made only with the degree
            of precision permitted by our measuring instruments. They are

more nearly exact now than in antiquity, when philosophers first dismissed measurement as useless to science. Thanks to the telescope, measurement is more precise than was possible for Tycho, who achieved as much accuracy as anyone could without the use of lenses. But even now, as we have learned, astronomical measurements may err by ten seconds of arc. Hence it is impossible to be sure that we always see *exactly* one-half the sky, in the sense required by philosophers for their pretended refutation of the earth's annual motion.

Sagredo   Very clever, and very efficient—except that it would take an hour to explain to such people how vast the effect of ten seconds of arc would be in this matter of enormous distances.

Salviati   Our friend does not try to explain that; he simply states the fact
GAW   and on that ground he denies the conclusion drawn by his adver-
417–18   sary. We do *not* see exactly half the sky at all times, he says, obliging his opponent to prove that we *do*—something that no one is yet capable of proving by any means whatsoever, or ever will prove by actual measurement.

Sarpi   I like this tactic very much, not only because of its efficiency, but because it is perhaps the only way of inducing such adversaries to pay attention to the basis of their assertions. In this case that would oblige them to master both practical mathematics and useful astronomy, concerning which they are so ready to pronounce without having any real idea of what they are talking about.

Sagredo   Our friend was scandalized a year or two ago when I wrote to him that I shared the common view which makes such neglect a definition of "philosophers," who nevertheless present themselves
GAW 190   as Nature's spokesmen and scribes. I daresay, however, that by the time he writes his book on the system of the world natural philosophers of the present age will cut a very poor figure in it.

Salviati   Certainly their recent attacks in matters of bodies in water and of sunspots have not elevated them in his opinion. But it is actual measurement, made as precisely as possible, that has chiefly altered his view of science. That made it clear that the kind of demonstrative proof hitherto demanded in natural philosophy is seldom possible, though an abundance of knowledge can be gained about Nature directly. I rather expect that when our friend writes on the system of the world he will do so not apodictically, as past astronomers and philosophers have done, but as does a sober judge, weighing concrete evidence and avoiding inveterate prejudices in this most admirable of problems.

Sarpi   Today's discussion has been so varied and so illuminating to me that I feel almost as if the book we have so long awaited had already been written and was now with the printers, where Salviati has skimmed through it to bring us word of its contents. Yet I know from what he told us yesterday that our friend has been occupied day and night in new observations, calculations, writings on other subjects, and replies to critics that have prevented his doing more than occasionally think about the organization of this vast, complex, and difficult work. You have done far more than enough to satisfy my original request, Salviati, and I stand much in your debt.

Salviati   Hardly more than I am indebted to you and to Sagredo, Fra Paolo. When we started talking this morning, I had only some hints about our friend's work done years ago at Padua, planting the seeds of a whole new science of motion. Now I see that that will probably lie at the basis of a principal part of his promised book, needed to clear away old objections against motion of the earth. The remarkable discovery of which Sagredo told us will point a way to connections between earthly and celestial motions, hitherto supposed nonexistent except by Kepler. Even Kepler would not grant the possibility of linking tides with motions of the earth, though from what I have learned today such a speculation was the beginning of our friend's interest in the new astronomy. So of the book that we have just, so to speak, been drafting for our friend to write, important parts would have remained unknown to me without your help. Perhaps most important of all will be that theory of the tides outlined in your notes of long ago, Fra Paolo.

Sarpi   I am delighted to have been able to contribute a little in return for the much I have learned from you and Sagredo, Salviati. I feel that perhaps we three, as a result of these conversations, are better able to anticipate what our friend's long-promised work will contain when it finally appears than anyone except the author himself. For that I thank you both.

Sagredo   Rather, let us all thank providence for having sent Salviati to Venice and then having brought you back here, Fra Paolo, before he continued on his way to Spain. And now, gentlemen, let us put aside speculation and enjoy the sights of Venice from our waiting gondola.

End of the Fourth Day

Epilogue

In 1632 Galileo published a book that he had spent five years organizing and writing, during which time he invariably referred to it as "my dialogues on the tides." That was not the printed title; in fact, technically speaking, the book had no printed title but the word "dialogue," followed by a very long subtitle. In 1635, after Galileo had been condemned by the Roman Inquisition for writing it, Galileo's *Dialogue* appeared in Latin translation with the title *Systema Cosmicum*; in 1661, in English translation it was called *The Systeme of the World*. Translators felt the need of a title and supplied one. Perhaps they chose as they did because they thought Galileo's book was the one he had promised in 1610. The purpose of this epilogue is to throw light on events after 1613 that resulted in Galileo's writing a dialogue on tides instead of a book on the system of the world. Deletion of his original title by Catholic licensers of printing has occasioned a variety of misconceptions about Galileo and his view of science.

The story up to 1613 was one of Galileo's expanding scientific knowledge. After that it was a story of changing tactics, partly voluntary and partly forced on Galileo, first by professors of philosophy and later by theologians. About the time the foregoing imaginary dialogue ended, philosophers began invoking the Bible against Galileo's belief in the earth's motion. In reply he challenged their credentials as scriptural interpreters. A year later his followers were denounced from the pulpit at Florence by a young priest. Rumors circulated that Copernican books would be prohibited by Catholic authorities at Rome. Galileo went to Rome late in 1615 to inform responsible theologians of accumulating evidence for the Copernican motions, lest action be taken that would deprive the Church, and Italy, of traditional leadership in science. Early in 1616 he wrote out, for a friendly cardinal, his explanation of tides. Neither that nor telescopic evidence dissuaded the authorities, who ruled that motion of the earth was foolish and absurd in philosophy and erroneous in the Catholic faith. An edict was issued regulating Copernican books, and Galileo was admonished not to hold or defend the Copernican motions any longer.

Having done all he could to circumvent a misstep that the Church need not have made, Galileo dropped the Copernican issue and occupied himself in other work for several years. In 1620 a Paris correspondent inquired about his book on the constitution of the universe, and Galileo replied that it had been stayed by a higher hand. As he wrote in 1623, he had been content to leave the public stage and communicate his thoughts only to friends until an unprovoked attack by a Jesuit professor induced him to reply on comets in particular and on scientific reasoning in general. That book, *The Assayer*, had nothing to do with planetary astronomy. It was dedicated to Urban VIII, the newly elected pope who as Maffeo Cardinal Barberini had long been a friend and admirer of Galileo's.

In 1624 Galileo visited Rome to pay homage to the new pope, who granted him six audiences during his stay. He returned to Florence bearing papal gifts and a warm letter of commendation to the Grand Duke. Urban was an intellectual, genuinely interested in science, who had appointed several members of the Lincean Academy to his staff. Though he did not rescind the 1616 edict, he said on a later occasion that if it had been up to him the edict would never have been issued. Galileo secured from Urban permission (and probably even encouragement) to proceed with the writing of what became the *Dialogue*, begun soon after his return to Florence. The circumstances were probably as follows.

While Galileo was in Rome, a German cardinal told Urban that the 1616 edict had to be handled with delicacy because German intellectuals were all Copernicans. The pope replied that the Church had never made Copernican views heretical, but only rash, being incapable of proof. Galileo knew of this conversation and also that Urban much wanted the support of intellectuals. He had explained his tide theory to Urban, who commented that it proved nothing, as God could produce the same effects in any number of ways. Now, a book written by a famed Catholic scientist with Church permission would be of value in countering opposition to the 1616 edict. That could be shown not to hinder scientific advance so long as the Copernican motions were treated only hypothetically and matters of biblical interpretation were avoided. Italian leadership in science, a tradition of which Urban was hardly less proud than Galileo, could also be made evident, while the edict would be seen to have been issued in full knowledge of all the scientific arguments, and for reasons of another kind. Such, I believe, was Galileo's proposal.

Galileo first composed a very long reply to the anti-Copernican arguments submitted in 1616 by the head of the organization for propagation of the Catholic faith. His tactics in this were to argue that not weak scientific arguments, but sound theological reasons had lain behind the edict. This essay was shown to the pope, who appeared to approve of it; it included the basic exposition of relativity of motion later used in the *Dialogue*. Next, Galileo started writing his "dialogue on the tides." The tactics here were to show that, apart from truth as revealed in the Bible, scientific explanation of tides could be given on the Copernican hypothesis.

Because it is now widely supposed that the 1616 edict prohibited Copernican books, it is necessary to explain the situation before 1633, when the Roman Inquisition first placed that official interpretation on it. The edict had in fact been carefully worded, probably by Robert Cardinal Bellarmine himself as personal theological adviser to Pope Paul V. It prohibited only two things: (1) attempted reconciliation of Copernican motions with the Bible and (2) positive assertion of their truth. Their hypothetical use for scientific purposes had been left open; in fact, Bellarmine had advised Galileo in 1615 to be content with such use, which sufficed for mathematicians and would not anger theologians. There was a long tradition behind Bellarmine's position, because the eccentrics and epicycles of Ptolemaic astronomy were generally regarded only as convenient fictions and not as literally existing. The official licensers of books at Rome and Florence, as also a special panel of theologians who examined the *Dialogue* after it was printed, interpreted the 1616 edict as permitting hypothetical use of the Copernican motions. In 1633 the Roman Inquisition ruled otherwise, but that does not affect prior events under discussion here.

Early in 1630 Galileo carried his manuscript dialogue on the tides to Rome for licensing, to be printed there as planned by the Lincean Academy. Responsibility for examining it was assigned to the Master of the Vatican, Niccolò Riccardi. After many required changes, intended to assure compliance with the 1616 edict, Riccardi reported to the pope that he would license the book. Urban, however, insisted that the words "on the tides" be dropped from its title. This requirement, which had far-reaching consequences, requires explanation; I believe the reason behind it was this.

Any value of Galileo's book to the Church hinged on its appearing with official permission, customarily printed on the title page. If the title remained "Dialogue on the Tides," it might appear that the Church endorsed a particular tide theory which assumed the Copernican motions. The simplest way to avoid any misconception was to delete the offending words from the title page.

Galileo went back to Florence to make the required alterations, intending to return to Rome for final approval. Soon afterward the head of the Lincean Academy died suddenly, leaving it without finances. Then an outbreak of plague closed the roads to Rome, and any manuscript permitted to pass would have to be fumigated page by page. Galileo applied to have the book licensed for printing at Florence, and in due course Riccardi wrote the following letter to the chief inquisitor there. Not only does it describe the *Dialogue* exactly as it was later printed, but Riccardi spoke authoritatively of the previous manuscript and of the reasons and conditions that made the book of value to the Church.

> Signor Galilei is thinking of publishing there a work of his that formerly had the title "On the Flow and Ebb of the Sea," in which probable reasoning is given concerning the Copernican system according to mobility of the earth, and he claims to facilitate understanding of that great arcanum of Nature with this position, reciprocally corroborating that by such utilization of it. He came here to Rome to show the work, which was subscribed by me, assuming the accommodations that had to be made in it and his bringing it back to receive final approval for printing. That being impossible to be done because of hindrances on the roads and danger to the originals, and the author desiring to finish the business there, Your Reverence may exercise your own authority and send the book forth or not, without any dependence on my review—but keeping in mind that it is the will of His Holiness that the title and subject may not propose the tides, but absolutely the mathematical consideration of the Copernican position about motion of the earth, to the end of proving that except for God's revelation and sacred doctrine, all the appearances could be saved in this position, collecting together all the contrary arguments that can be adduced from experience and the Peripatetic philosophy, in such a way that absolute truth is never conceded to this position, but only hypothetical, and without the Bible. It must also be shown that this work is done solely to show that all the reasons are known, and that it was not for lack of knowledge that this opinion was banned at Rome, in accordance with the beginning and ending of the book which I shall send, adjusted, from here. With this precaution the book will have no impediment here at Rome, and Your Reverence will be able to satisfy the author and to serve His Serene Highness who is showing so much pressure in this. Remember me as your servant, and favor me with your commands.

The pope's removal of the tides as title and subject of Galileo's *Dialogue* created a problem for the author because his entire organization and composition of the book over a period of five years had been based on the original title. The adjusted beginning of the book sent by Riccardi consisted only of the author's preface to his readers, and it is not hard to see that the adjustment consisted mainly, if not entirely, in transposing Galileo's statements about tides from the beginning to third place. The original beginning speeches of the dialogue itself must have concerned the problem of explaining the tides, as seen from a passage much later in the book, and as would be consistent with the last sentence of the preface. Removal of those opening speeches (and reordering of the preface) left the printed book without any clear organizational principle, so Galileo's discussion of tides at the end has appeared to modern commentators as a disconnected afterthought clumsily added at the end. It has even been called an improper and forbidden attempt to supply demonstrative proof of the earth's motions, though Riccardi himself had approved it as "probable reasoning," which never has constituted demonstrative proof.

The shift in the concept of science during the seventeenth century, already described in the introduction, is overlooked in the modern confusion of demonstrative proof with probable reasoning. Like Galileo, scientists today proceed mainly on great preponderance of evidence. In the old philosophical concept of science that could never constitute more than probable reasoning. Demonstrative proof required rigorous logical deduction from metaphysical first principles. Hence the meaning of "scientific demonstration" was quite different before Galileo from what it is now. Appeal only to actual measurements would not then have constituted demonstrative proof in science. That is hardly the case in physical science today. The strength of Galileo's case for motion of the earth lay in the weakness of metaphysical reasons against it, not in anything that in his day would have been regarded as demonstrative proof by anyone, least of all by Galileo.

Science was presented in the *Dialogue* as a never-ending inquiry into phenomena of nature. Whether that was a tactical device adopted for compliance with the 1616 edict, or Galileo's own view of the matter, is debatable. Perhaps it began as the former and ended as the latter.

Appendix
Galileo's Diagrams and Measurements

Drawings made by Galileo for use in printing *Sidereus Nuncius* illustrations of satellite positions. Reduced facsimiles from Edizione Nazionale of Galileo's *Opere*; concerning his scale at top, see ensuing *Discussion*.

January 25–February 11, 1610

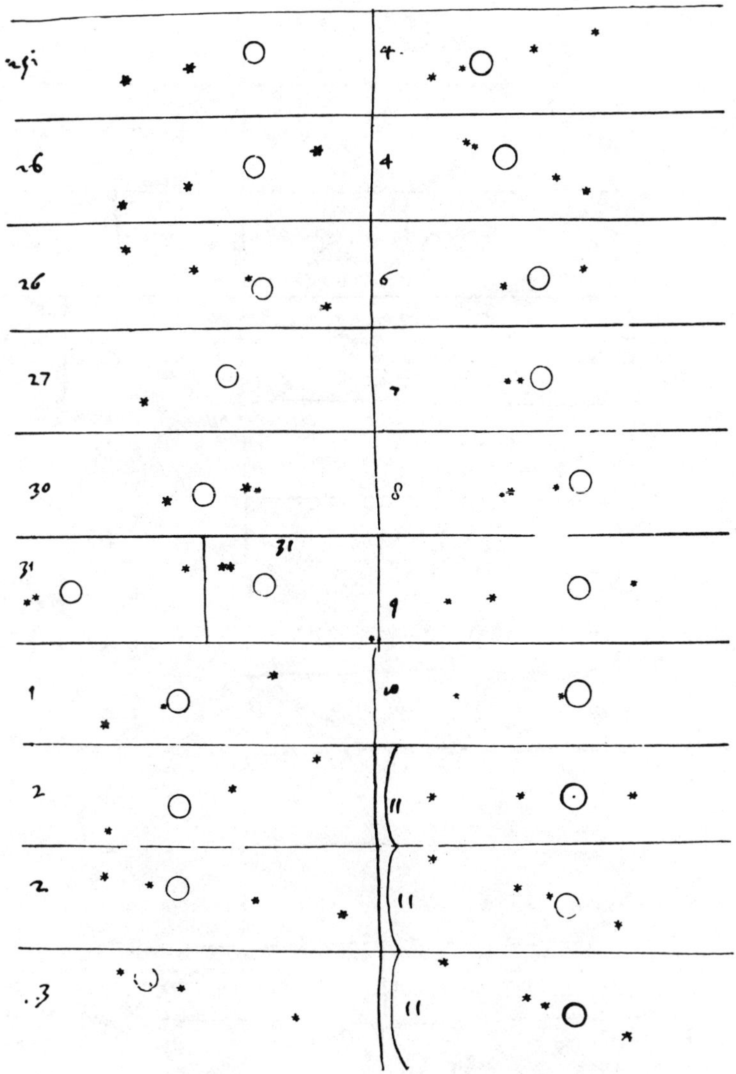

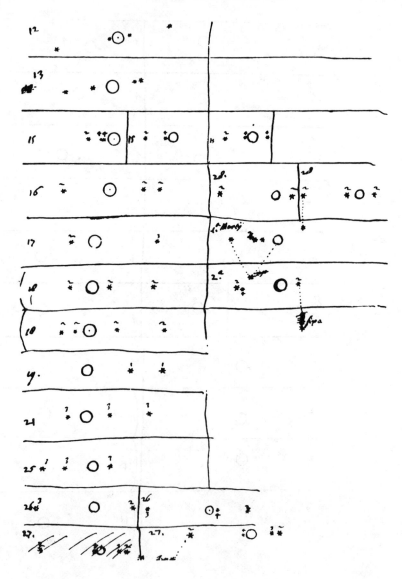

Discussion of Galileo's Diagrams and Measurements

At the top of the first page of Galileo's set of drawings for diagrams of the positions of Jupiter's satellites he drew a scale but did not identify the units. If his unit was one "minute," as that term appears in the text of the *Starry Messenger*, the drawings conform pretty closely with Galileo's verbal descriptions of observed positions. On that basis, however, the diameter of Jupiter as drawn was two "minutes." One must conclude that the disk of the planet was not drawn to the same scale as the separations between satellites, or between the edge of Jupiter and the nearest satellites. The greatest separation described is 14 "minutes," implying elongation from Jupiter's center of no more than 14½, whereas true maximum elongation is 26½ Jovian *radii*. The "minute" was probably Galileo's estimate in Jovian *diameters*, his drawn diameter being made double the corresponding size, perhaps by reason of the difficulty of setting an ink compass to draw very small circles neatly. How Galileo succeeded as well as he did in drawing fine ink lines with a quill pen is a mystery to me. The paper shows punctures by the point of Galileo's compass at the center of Jupiter; yet even today, neat small circles are difficult to execute in ink with steel pen compasses.

Early in 1612 Galileo began using 24 radii as the radius of the largest satellite orbit. At that time it is certain that his telescope showed Jupiter's disk with a spurious ring of light that extended the true radius 10 percent. From his earliest carefully drawn "jovilabe" it is clear that he began, probably in 1611, by taking the radius of the largest orbit to be 15 "minutes." Assuming the same spurious ring of light and taking as R the radius seen by Galileo, r being the actual radius, we would have $23R = 14$ "minutes," and $R = 0.6$ "minutes" or 35 "seconds" approximately. This confirms the probability that when he wrote the *Starry Messenger* Galileo took the apparent diameter of Jupiter to be one "minute," and overcounted considerably when he observed a maximum elongation of the outermost satellite with no intermediate satellite present from which he could count.

If the central circle in the drawings for the earliest jovilabes is supposed to represent the disk of Jupiter, this too is double the size it would be if it were drawn to scale. Very likely the central circle at first simply represented a radius of one "minute." Beginning early in 1612 Galileo represented the apparent disk of Jupiter by a central circle drawn to scale, abandoned "minutes," and reported everything in apparent radii of Jupiter. The circles he drew for satellite orbits about March 1612 measure $5.5R$, $9R$, $14R$, $24R$; modern values are $5.9r$, $9.4r$, $15r$, $26.4r$.

The following table shows first the calculated positions of the satellites with respect to Jupiter as measured in radii from the center of the planet. Next are shown the positions obtained by using the scale drawn by Galileo at the top of his set of drawings to measure distances shown in those drawings. Finally, the positions implied by Galileo's descriptions in "minutes" in the text of his book are shown, taking that word to mean apparent diameters of Jupiter. One radius, or one-half "minute," was added to those descriptions which applied to separations from the edge of Jupiter, and conversion to radii was made by doubling. A single dash indicates that a satellite lay within Jupiter's disk in the column designated "Actual," or that Galileo reported no satellite in the other columns. Dashes before and after an entry indicate that two satellites were seen as one by Galileo (usually when their separation did not exceed one-half radius of Jupiter, or about ten arc-seconds).

Elongations of Jupiter's Satellites in Jovian Radii

Obser- vation Num- ber	Actual				Measured Diagrams				Implied in Sidereus Nuncius			
	I	II	III	IV	I	II	III	IV	I	II	III	IV
1	E6	E6½	W14	E24	-E4-		W9	E14				
2	W5	W8½	W12½	E26½	W7	W8	W13	-				
3	W2	E9½	E10	E20½	-	-E6-		E12				
4	W3½	W3	E15	E11½	-	-	E15	E10				
5	E3	W9	E9	E3	-	W5	E7	-		W5	E5	
6	E4½	W8½	E8	E2½	E1½	W5	E7	-	E1½			
7	W5½	E5	W4	W7	W3½	E5	W1½	W5½	W5	E5	W3	W7
8	W5	W8	W13	W23	W5	W9	W13	W21	W5	W9	W13	W21
9	W2½	W9	W12	W23½	-	W10	W12	W24		W7	W9	W18
10	-	W9	W11½	"		IR apart						
11	E4½	W4½	W4	W26	E2	-W2-		W17	E2+	-W2+-		W17
12	W3½	E8½	E8	W26	-	-E5-		W17		-E7-		W23
13	-	E9½	E10+	W25½	-	E6	E8	W20	0.6R apart			
14	E1½	E1½	E15	W22	-	-	E16	W20			E17	W21
15	E2½	W9½	E10	W15	-	W9	E11	W17		W11	E13	W19
16	E4				E2	W8	E11	W14				
17	E4½	W8	E9	W14½	E5	W10	E12	W16				
18	E5	W8½	E8½	W14								
19	W4	E2½	W1	W6½	W3	E2	-	W5	W3	E3		W5
20	W6	E6	W4½	W4½	W4	E4	-W3-		W3	E5	-W2.6-	
21	W6	E6½	W5	W4	W4	E4	W3	W2	W3	E5	W2+	W2-
22	E5	E8½	W12	E3	E3½	E5½	W8	E2	E5+	E8	W9	E3-
23	W6	W6½	W14	E13	W3	W5	W14	E11	W3	W4-	W15	E11
24	W4½	W8½	W13½	E14½	W3	W6	W9	E11	W4-	W6+	W9	E11
25	E5½	W6	W6	E20½	E4	-W5-		E16	E5+	-W8-		E19+
26	E3½	W3½	W3½	E21½	-	-	-	E14				"
27	W4	E8½	E7	E25	-	E7	E5	E21		E6	E5	E24
28	-	E9½	E9½	E25½	-	-E7-		E21½		-E7-		E23
29	E1½	E2	E15	E26½	-	-	E11	E23			E13	E23
30	-	W9½	E12	E24	-	W10	E11½	E23½		W11	E12-	E24-
31	E3½	W9	E10	E23	E2	W11	E12	E25				
32	W2½	E1½	-	E18	-	-	-	E15				E15
33	E6	W7	W7½	W10	E6	-W6½-		W8½	E6	-W7-		W9
34	W4½	E8½	E6	W18½	-	E7	E5	W20		E7-	E5+	W21

35	W3+	E9	E6½	W19	-	E7	E5	W20				
36	E1½	E3½	E14½	W24	-	E1½	E13	W17		E2−	E13	E17
37	-	W9½	E12½	W26½	-	W9	E13	W25		W9	E13	W25
38	E5	W9	E10½	W26½	E4	W13	E12	W29	E4+	W13	E12½	W29
39	W6	E4½	W1½	W24	W5	E4	-	W27	W5	E4		W25
40	E5	E7	W10½	W20	E3	E9	W9	W21	E2+	E8+	W9	W21
41	E6	E8	W12½	W18½	E6	E8	W9	W15	E5	E6	W9	W15
42	E5	W6½	W7½	W2	E5	-W7-		-	E5	-W7-		
43	W3½	E8	E5	E8	-	E5	E3	(E5)		E7	E5	(E7)
44	E3½	E6½	E13½	E15½	E3−	E3	E11	E13	E3?	E4−	E12−	E12+
45	E2	E5	E14	E16½	-	E3	E12	E13+	E1−	E3	E12	E13+
46	E1½	E4½	E14	E16½								
47	W2	W8	E14	E22½	-	W9	E15	E23		W9	E15	E23
48	W1½	W3	E4	E26	-	-	E1½	E21			E2−	E21
49	W3½	W1½	E2½	E26								E25
50	E3	E9½	W8½	E26	-	E8	W9	E24		E9	W9	E25
51	E4½	E9½	W9½	E25½	E2	E9	W9	E25				
52	E5½	E9	W10½	E25½	E4	E8	W9	E24				
53	W4½	W1½	W15	E22½	W3	-	W17	E21	W3		W17	E21
54	W5	W4	W15	E21½	Mistakenly reported star no longer seen							
55	E5½	W9	W11	W15½	E5	W7	W9	E15	E5	W8	W9	E13
56	E4½	E6½	E12½	W3½	E2	E3	E9	-	E3−	E3+	E7+	
57	E2	E4½	E13½	W5	-	E3	E9	-		E2	E9	
58	E1	E3½	E13½	W5½	-	E3	E9	W5				W5
59	E1½	W9	E13½	W14½	-	W11	E15	W17		W11	E15	W17
60	-	W3½	E6	W20	-	-	E7	E21			E7	E21
61	W3½	-	E3½	W21½							E3−	W21
62	E1½	E9½	W6½	W25	-	E7	W5	W21		E7	W5	W21
63	E2	E9	W8½	W25½						E7	W7	(W21)
64	E5	E9	W9	W25½	E3½	E8½	W9	W26	E5−	W11−	W7	W21
65	W3½	-	W14½	W26½	-	-	W15	W27			W15	W27
66	W6	E4½	-	W18	W7	E5	-	W21	W7	E5		W21
67	-	E9	W5	E18	-	E9	W5	E17		E9	W5	E17
68	W2½	E1½	W14	E24	-	-	W13	E21			W13	E21
69	W5	W1½	W14½	E24½	W3	-	W13	E22	W3		W13	E23
70	E5	W9½	W13	E26½	E2	W7	W10	E21	E2	W6	W8	E21
71	E6	W9	W12	E26½					E3	W6	W8	E21
72	W5½	E2½	W2½	E25	W5	-	-	E19	W5			E19
73	W5½	E5	-	E24½	W5	E4	-	E20	W5	E5		E19
74	E6	E8½	E9½	E20½	E5	E8	E10	E17	E5	E7	E8	E16
75	W5	W5½	E15	E13½	-W5-		E15	E13	-W5-		E15	E14

Notes

1. Galileo's working papers on motion show that Galileo set them aside in 1610 and did not resume those studies until 1618; see *GAW*, 262–63 and my monograph *Galileo's Notes on Motion* (Florence: Istituto e Museo di Storia della Scienza, 1979).

2. Sagredo was absent from Venice in 1606, when he served as treasurer in Palma, Majorca, and again from August 1608 to July 1611, when he was the Venetian consul at Aleppo, Syria.

3. *Ambassades du Roy de Siam a l'Excellence du Prince Maurice*, printed anonymously at The Hague in October 1608. A facsimile was published with S. Drake, *The Unsung Journalist and the Origin of the Telescope* (Los Angeles: Zeitlin & Ver Brugge, 1976).

4. John Donne, *Ignatius His Conclave* (London, 1611).

5. The sun was taken as the center of planetary motions not only by Copernicus but by Tycho Brahe, though the latter denied that the earth was a planet. Hence Galileo's statement here is not decisive as to Copernicanism on his part.

6. Aristarchus, a century before Hipparchus, had put forth the hypothesis of motions of the earth. Ptolemy noted the utility of the heliocentric assumption for some planetary phenomena but did not pursue it. Medieval philosophers discussed

possible rotation of the earth but not its revolution. The Copernican theory entailed the first serious threat to Aristotelian physics, only temporarily averted by the Tychonic compromise.

7. Salviati's allusion to a remark of Galileo's is fictional and will be amplified in the Fourth Day. It is probable that not even Sagredo knew Galileo's explanation of tides before 1616; see *Opere* xii, 287–88. Here it is assumed that he did.

8. In Galileo's surviving correspondence from 1598 to 1610 there is no mention of Copernican astronomy apart from one unanswered inquiry sent by Kepler to an intermediary (Edmund Bruce) at Padua; see *Opere* x, 75, and *GS* 128–29.

9. Two telescopes and one cracked objective lens are preserved as Galileo's at the Museum of Science History at Florence. The less powerful has wooden tubes, covered with reddish paper; it magnifies 14× and shows streaks of red and blue radially around a bright image. The other, which enlarges 20×, is cardboard covered with red leather ornamented in gold; its image of a bright light is not round, but elongated. Using a telescope made by Galileo, Kepler noted both chromatic aberration and oblong image (*Opere* iii, 185). The cracked objective lens was mounted in its engraved ivory mounting in 1677, the damage having occurred during Galileo's lifetime. When used with the eyepiece of the weaker telescope, it gives magnification of 18×. Galileo sent to Cosimo with the dedication copy of *Sidereus Nuncius* "that same spyglass with which I discovered the planets and made all the other observations. I send it unadorned and rough, as I made it for my own use; but since it has been the instrument of so great a discovery, I wish it to be left in its original state" (*Opere* x, 297). This suggests that the telescope covered with tooled leather may have been sent previously to Cosimo, it being of sufficient power for observing the Medicean stars, though of mediocre quality. Most experts agree that the relics are authentic, except that the eyepiece of the 20× telescope may be of later date.

10. Sarpi described this (*Opere* x, 290) as 4 Venetian feet in length, with a concave eyepiece ground spherically to a radius of less than a finger's breadth and an objective ground to a radius of 6 feet. With it, he said, an object subtending 6 arc-minutes was seen as occupying 3 degrees of arc. The focal length of the broken objective lens mentioned in the preceding note is 1.7 meters. At the time of metric conversion in Italy, the Florentine foot was about 13 inches; if the Venetian foot was near that, in 1610, Sarpi's description would fit the broken objective fairly well, since he probably judged the radius by the focal length.

11. Since there is little doubt that Galileo meant by "minute" the visual diameter of Jupiter, his stated range of accuracy for observations recorded early in 1610 was about right. Yet the method of measurement proposed was impracticable. Perforated plates literally "fitted to the lens" would not reduce the field of view appreciably; to do that, they would have to be placed well beyond the lens. As Galileo admitted in 1612, speaking of these earlier records: "I had not found a way to measure with any instrument the distances between the planets, so that I made notes of their separations merely in terms of the diameter of Jupiter judged by eye, as we say" (*CES*, 19).

The micrometric instrument he devised early in 1612 will be described in the Third Day.

12. This is what Galileo said to the French canon Jean Tarde, who visited him in November 1614. Tarde wanted instructions for making fine telescopes, which at that time Galileo was unwilling to give to foreigners. Tarde's subsequent behavior proved Galileo's mistrust of him to have been justified; see *GS*, 190.

13. The large spot meant is *not* the circular one discussed below but an irregular area near the top, crossing the boundary of light at first and last quarters as shown.

14. The contour in black invading the lighted portion here is exaggerated in blackness and agrees poorly with the white contour in the second diagram, which repeats that for last quarter previously shown.

15. The circular crater, shown disproportionately large in the woodcut, is now called Albategnius. Had the lunar names been assigned by friendly astronomers, it would certainly be known as Galileo, the name given by Jesuit selenographers to a tiny and undistinguished crater. Galileo's diagrams of the circular crater are discussed by Righini, Gingerich, and Whitaker in recent papers cited in the Preface.

16. Fixed stars are not magnified at all; they appear brighter by reason of the light-gathering power of telescopes. Nevertheless, Galileo's discussion corrected an enormous error in traditional estimates of sizes of fixed stars (and planets); see further in *D*, 335–39 and 359–64.

17. Galileo's maps of constellations, here and in his notes, show that he detected stars of seventh and eight magnitudes, and some of ninth. See also Drake and Kowal, "Galileo's Sighting of Neptune," *Scientific American* (Dec. 1980): 74–78.

18. Aristotle took the Milky Way to be a sort of high fog in the elemental region, just below the lunar sphere. Arab astronomers remarked on its lack of parallax, leading to controversy over Aristotle's meaning and explanation of the facts. Most natural philosophers came to regard the Milky Way as a dense part of the aether reflecting light; compare Galileo's discussion of nebulae, below.

19. Galileo's original notes make it probable that he did not draw this conclusion until the night of January 12, when he again saw three satellites, and indeed not until his second observation on that night; see *GAW*, 150. The idea of stars circling anything but the earth, or possibly the sun, was simply unthinkable at the time.

20. Four satellites had been visible on 8 January, but one of them was far to the east of Jupiter, while the other three were closely grouped to the west. Simon Mayr, who in 1614 claimed to have been observing the satellites since late in November 1609, but not to have recorded them until 8 January 1610 (29 December 1609 by the Julian calendar), made the identical oversight, if he is to be believed. Mayr did not recognize the existence of four satellites until late February or early March. His first authenticated observation was on 30 December 1610, exactly one year after his alleged earliest recorded position.

21. Concerning the measuring unit and scale employed by Galileo when writing the *Starry Messenger*, see Appendix. In the diagrams, Jupiter is drawn double the size

to be expected if its diameter was treated as the "minute" used in measuring separations. The next diagram and its description illustrate the puzzle.

22. Now named Callisto, the others in succession being called Ganymede, Europa, and Io after four mythical attendants of Jupiter. The names were suggested to Simon Mayr by Kepler and were not published until 1614, the year after the discussion here.

23. Galileo's drawings included the diagram for (16), which lacks a counterpart in the 1610 edition. The diagram for (17) was not included among Galileo's drawings but is found in the printed text of 1610.

24. The smallest distance at which Galileo could distinguish two bright points in a dark field was in fact about ten arc-seconds, the resolving power of the 20× and 18× telescopes described in note 9, above. A satellite near the bright disk of Jupiter was seldom recorded by Galileo at even one arc-minute from the true edge of the planet, and until 1612 he almost always greatly underestimated such separations.

25. The error apparently occurred through Galileo's taking a new count for separation of the nearer westerly satellite from Jupiter, but not again counting its separation from the farther westerly satellite, previously recorded as eight "minutes." It happened that the place of the latter, which was IV, was of particular importance on this night, as will be explained later in the discussion. Only (38) recorded for it an impossibly great separation from Jupiter.

26. It was very difficult for Galileo to count by eye through large separations from Jupiter with no intermediate satellite. Actual motion of the westerly satellite (again IV) had been much less than the two "minutes" here implied during five hours.

27. Doubtless Galileo had in mind the sun as that center, but since he did not say so, his statement here was acceptable in any of the rival astronomies.

28. Up to this point, Galileo merely described the Copernican assumption without committing himself. Whether the balance of the paragraph committed him is an interesting question for logicians. If the phrase "all together" includes the earth and moon, Galileo spoke as a Copernican; if it means only Jupiter and the four satellites, there is still no commitment on his part beyond the implication inherent in the Copernican assumptions as such, that the earth is a planet.

29. It is an interesting fact that those who published attacks against the *Starry Messenger* did not single out Copernican implications as blameworthy. Not motion of the earth, but roughness of the moon and existence of new moving stars were most energetically opposed at first. Only when defeat of Aristotelian natural philosophy on those matters was certain did philosophers seek to have Galileo silenced by the Church as a Copernican.

30. The satellite orbits are in fact very nearly circular and are much too small to account for great changes in brightness at different positions of the satellites.

31. As Galileo had written in his *Sunspot Letters*: "Even the most trivial error is charged to me as a capital fault by the enemies of innovation, making it seem better to remain with the herd in error than to stand alone in reasoning correctly" (*D&O*, 90).

32. Plato sketched somewhat cryptically in the *Timaeus* an account of celestial

motions and is said to have proposed as a problem for others the determination of uniform circular motions about the earth as center which could account for observed irregularities of uniformity so established. Eudoxus worked out the appropriate scheme of homocentric spheres, without bothering about actual measurements. Aristotle, in his *Metaphysics*, supplied additional spheres required not by mathematics, and still less by astronomical measurements, to provide a causal explanation of all celestial motions in terms of a single motive power.

33. Galileo began indicating estimated magnitudes by numbers on his drawings on 15 February 1610, as may be seen from the reduced facsimiles of his drawings for the *Starry Messenger* (see Appendix). The observations recorded in December 1610, about to be discussed, are shown below, from *Opere* iii, p. 439.

34. The same lunar eclipse was observed by Simon Mayr, who in 1614 claimed priority of about one month over Galileo in discovery of Jupiter's satellites. Only half a dozen dated and described observations are found in all of Mayr's publications or in letters of his friends. The earliest that can be authenticated was on the night of this eclipse, 29/30 December 1610, or 19/20 December in the Julian calendar used by Mayr. The earliest he claimed to have recorded, after a month of casual observations, was on 8 January 1610, or 29 December 1609 in the Julian calendar.

35. Simon Mayr stated in 1614 that he had found only the periods of III and IV by March 1611; in April–May of that year, Kepler thought the period of III to be about 8 days, nearly a day too long.

36. The Copernican and Ptolemaic corrections differed nearly one degree at the maximum, making it possible to identify the figures Galileo used first (1611 and 1612) as taken from Ptolemaic tables and those he used later (in 1613) as Copernican.

37. Galileo first mentioned this instrument in a note on 31 January 1612. For a description and illustration of it, see Drake and Kowal, "Galileo's Sighting of Neptune."

38. An earlier undated calculation (*Opere* iii, 859) placed the diameter of Jupiter at 50 arc-seconds. For 21 January 1612 Galileo calculated it to be $41''.6$, and for 9 June 1612, $39''.4$, Jupiter having become more distant from the earth. Measurements to a few seconds of arc had never been possible before, and these were commendably accurate.

39. The evolution of the jovilabe appears not to have been studied, though its surviving examples are reproduced in *Opere* iii, 477–87. The first draft, p. 481, was probably drawn about June 1611, followed by p. 483 about October, and then p. 477 near the end of 1611. For these, maximum elongation of IV was put at 15 "minutes." After the drawing on p. 479, Galileo probably returned to p. 477 around early February 1612, adding new graduations in Jovian radii out to 24 for IV. Soon afterward he drew the definitive paper jovilabe of pp. 486–87, which bears signs of frequent use. The unique brass jovilabe preserved at Florence dates probably from 1617.

40. Giacomo Zabarella, ranking professor of philosophy at Padua when he died in 1589. His successor was Cesare Cremonini (the Simplicio of *CES*), who continued Zabarella's enlightened Paduan Aristotelianism by firmly rejecting mathematics in natural philosophy; see *GAP*.

41. Sarpi's notes on philosophical and scientific questions survive in a manuscript copy at the Marciana Library in Venice. Written in numbered paragraphs over a period of thirty years, they do not evidence Copernican leanings by Sarpi. In Micanzio's biography of Sarpi he was said to have formulated a tide theory utilizing but a single motion of the earth, in contrast with the theory summarized in the four paragraphs to be cited below. The latter, elaborated by Galileo in the Fourth Day of his *Dialogue*, is nowhere else mentioned in Sarpi's notebooks.

42. For details see my "Ptolemy, Galileo, and Scientific Method," *Studies in History and Philosophy of Science* 9 (1978): 99–116.

43. Science, for Aristotle, required knowledge of causes, attainable only by reason. Knowledge of facts acquired by experience was insufficient to constitute a science. When astronomers agreed not to discuss causes of celestial motions, Aristotelians could properly maintain the view here imputed to La Galla.

44. Traditionally, Pythagoras was led to his mathematical theory of music by noting numerical ratios among the weights of hammers ringing in consonance after striking an anvil. Galileo's father, Vincenzio Galilei, discovered by careful measurements that the Pythagorean rules could not be extended from lengths of musical strings to string tensions, as had been previously assumed in music theory, in which he pioneered a great revolution.

45. No document supports this account, by which the presence in Sarpi's notebooks of a tide theory that was not his (see note 41) is plausibly explained.

46. Probably Christopher Wursteisen, who enrolled in law at the University of Padua in 1595. In Galileo's *Dialogue* the name was given as Christian Wursteisen, who never visited Italy but who published a book in which Copernican astronomy was mentioned. See *GAW*, p. 36.

47. This paragraph (568) was first published in *GS*, p. 201. Those that follow will be found in the abridged text of Sarpi's notebooks, *Scritti Filosofici e Teologici*, edited by R. Amerio (Bari, 1951), p. 115.

48. The "third motion" of Copernicus to explain the maintenance of the earth's axis always nearly parallel to itself was shown by Galileo to be not properly a motion at all, but a consequence of the relativity of motion; cf. *D&O*, 264–65; *D*, 263.

49. Galileo wrote out a long elaboration of his tide theory in 1616, but had still not managed to relate cyclical variations in height of tides to phases of the moon. His ultimate solution of that problem, mistaken but ingenious, was set forth only in 1632; see *D*, pp. 452–54. Contrary to the prevailing view that Galileo believed planetary motions to be uniform and circular, he invoked for spring and neap tides nonuniform motion of the earth in its annual orbit, related to lunar positions.

50. It is probable that Galileo did not make the connection until early in 1609, whereas Sagredo left Venice in August 1608. Evidence that he had made the connection before returning to Florence in 1610 exists in his notes on motion and the same is implied in a letter describing his planned book on the system of the world in 1610. Sagredo is here supposed to know of this "Platonic cosmogony."

51. In order to make their calculations, Ptolemy and other early astronomers needed to assume uniform circular motion of each planet around some point, not necessarily the same point for all planets. This point was called the *equant*. Copernicus regarded these empty points as a sophistical device, not truly satisfying the traditional requirement of uniform circular motions around the center of the universe. "If you say that the epicycle moves regularly around the center of the

earth, . . . then what sort of regularity will that be which occurs in a circle foreign to the epicycle . . . ? They mean the regularity of the moon in its epicycle to be understood not in relation to the center of the earth, . . . but in relation to some other different point, which has the earth midway between it and the center of the eccentric circle" (*De revolutionibus*, IV, 2).

52. The necessary assumptions turned out to be remarkably simple, at least in appearance, when they were finally found by Newton, using mathematical tools not available to Kepler or Galileo.

Index